Video Techniques in Animal Ecology and Behaviour

Video Techniques in Animal Ecology and Behaviour

Edited by

Stephen D. Wratten

Department of Entomology and Animal Ecology
Lincoln University
New Zealand

 CHAPMAN & HALL
London · Glasgow · New York · Tokyo · Melbourne · Madras

Published by Chapman & Hall, 2–6 Boundary Row, London SE1 8HN

Chapman & Hall, 2–6 Boundary Row, London SE1 8HN, UK

Blackie Academic & Professional, Wester Cleddens Road, Bishopbriggs, Glasgow G64 2NZ, UK

Chapman & Hall Inc., One Penn Plaza, 41st Floor, New York NY10119, USA

Chapman & Hall Japan, Thomson Publishing Japan, Hirakawacho Nemoto Building, 6F, 1-7-11 Hirakawa-cho, Chiyoda-ku, Tokyo 102, Japan

Chapman & Hall Australia, Thomas Nelson Australia, 102 Dodds Street, South Melbourne, Victoria 3205, Australia

Chapman & Hall India, R. Seshadri, 32 Second Main Road, CIT East, Madras 600 035, India

First edition 1994

© 1994 Chapman & Hall

Typeset in Times by Interprint Ltd. Malta
Printed in Great Britain by St Edmundsbury Press, Bury St Edmunds, Suffolk.

ISBN 0 412 46640 6

A catalogue record for this book is available from the British Library

Library of Congress Cataloging-in-Publication data

Video techniques in animal ecology and behaviour/edited by Stephen
D. Wratten. — 1st ed.
    p.   cm.
    Includes bibliographical references and index.
    ISBN 0-412-46640-6 (acid-free paper)
    1. Nature photography. 2. Video recording. 3. Animal ecology—
Research. 4. Animal behavior—Research. I. Wratten, Stephen D.
TR721.V53   1993
591.5′072—dc20   93-32960
                                            CIP

♾ Printed on permanent acid-free text paper, manufactured in accordance with the proposed ANSI/NISO Z 39.48-199X and ANSI Z 39.48-1984

Contents

Contents   vii

Contributors

Dr Stuart Bailey
Department of Environmental Biology
Williamson Building
School of Biological Sciences
University of Manchester
Oxford Road
Manchester M13 9PL   Fax (+44)61 275 3938

Dr Mike H. Bowie
Department of Entomology and Animal Ecology
PO Box 84
Lincoln University
Canterbury
New Zealand   Fax (+64)3 325 3844

Dr John Bradshaw
Anthrozoology Institute
Department of Biology
University of Southampton
Southampton S09 3TU   Fax (+44)703 558163

Dr Mike Copland
Department of Biological Sciences
Wye College
University of London
Wye, Ashford
Kent TN25 5AH   Fax (+44)233 813140

Dr Gabriella Gibson
Department of Biology
Imperial College at Silwood Park
Ascot
Berks SL5 7PY   Fax (+44)344 294339

Mr Chris D. Hall
Scottish Office Agriculture and Fisheries Department
Marine Laboratory
PO Box 101
Aberdeen AB9 8DB   Fax (+44)224 295511

Dr Jim Hardie
Department of Biology
Imperial College at Silwood Park
Ascot
Berks SL5 7PY   Fax (+44)344 294339

Dr Helen Nott
Waltham Centre for Pet Nutrition
Waltham-on-the-Wolds
Melton Mowbray
Leicester LE14 4RT   Fax (+44)664 415440

Dr Joe Riley
National Resources Institute
Radar Unit
RSRE
Leigh Sinton Road
Great Malvern
Worcestershire WR14 1LL   Fax (+44)684 582984

Dr Chris Sherwin
University of Bristol
Department of Animal Husbandry
Langford House
Langford
Bristol BS18 7DU   Fax (+44)934 853145

Dr Ken W. Smith
RSPB
The Lodge
Sandy
Bedfordshire SG19 2DL   Fax (+44)767 692365

Dr Kazuyuki Sugino
Institute of Basic Medical Sciences
University of Tsukuba
Tennodai
Tsukuba
305 Japan   Fax (+81)298 53 3098

Mr Jon Varley
Department of Biological Sciences
Wye College
University of London
Wye, Ashford
Kent TN25 5AH   Fax (+44)233 813140

Dr Clement Wardle
Scottish Office Agriculture and Fisheries Department
Marine Laboratory
PO Box 101
Aberdeen AB9 8DB   Fax (+44)224 295511

Dr Steve D. Wratten
Department of Entomology and Animal Ecology
PO Box 84
Lincoln University
Canterbury
New Zealand   Fax (+64)3 325 3844

Dr Stephen Young
Department of Biology
Imperial College at Silwood Park
Ascot
Berks SL5 7PY   Fax (+44)344 294339

Preface

This book is about video techniques, not video technology. To deal with the latter would be a thankless task, as by the time the book was published it would be out of date, given the rapid rate of development of video hardware. However, these technological advances do help to make it an exciting field. As Joe Riley says in Chapter 1, 'Advances in video technology continually produce improvements in performance and reductions in both the cost and size of equipment, so it seems certain that the technique will prove to be an even more useful resource ... in the future.'

In selecting the topics and authors represented in this book, I have tried to encompass most of the behavioural and ecological uses to which video is likely to be put over the next decade or so. I believe the book has captured the invaluable accumulated experience of the most active practitioners of the medium in this research area. The idea for the book arose from two workshops on the use of video in ecology and behaviour held at Southampton University, UK, during the 1980s. Three learned societies were involved in these meetings: the Association of Applied Biologists, the British Ecological Society and the Society for Experimental Biology. Grants from the UK Natural Environment Research Council and the Royal Society of London enabled Dr Nigel Halsall and me to become involved in developing video for the analysis of invertebrate predation, and therefore, together with the aforementioned meetings, led indirectly to this book. Drs Bob Carling and Clem Earle at Chapman and Hall kept the momentum going with their enthusiasm, and my co-authors worked hard to meet deadlines during their busy research schedules and despite the disruption of my leaving, like Tennyson's Ulysses, ' ... to seek a newer world' during the book's production.

In these pages, you will be guided around the frustrating delays and pitfalls associated with: monochrome or colour; time-lapse settings; choice of background; 3-D video; PAL or NTSC; flare; illumination; wavelengths for illumination; spectral sensitivity of video cameras; image size; split-screen images; solid state (CCD) cameras; time–date generators; 'judder' in pause mode; use of cameras in field conditions, including under water; low-light images; choice of lenses; digitizing of images; automatic and semi-automatic tracking; RGB video format; choice of arenas; data extraction; genlocks, timebase corrections; video analysis software and much more.

When embarking on a new research medium, whether it be DNA technology or the use of research video, there is nothing better than to talk to a friendly, communicative expert; a colleague in the BBC once described to me the pleasures of the 'talk' programme, Radio 4, as like those of ' ... spending time with an erudite friend'. I hope this book provides the same type of erudition and support to those embarking on video to help them answer questions in behavioral ecology; video is certainly a research tool which is increasing rapidly in its contribution to published work in this area, as Chris Sherwin shows in Chapter 7. However, although the number of studies using video in his speciality, that of the behaviour of farm animals, has increased, as a proportion of total studies reported in a major journal there has been no upward trend. Chris asks whether this is because researchers consider that video has nothing new to offer or because they are ignorant of the techniques available and of the advantages of the medium. I believe the latter explanation is the more likely, and hope that this book will convince biologists on the brink of introducing video into their research protocols that it has a lot to offer. As 'Bill' Bailey points out in Chapter 4, although there are no equipment costs involved in the alternative to video – direct observation – the 'downside' is ' ... extreme boredom, loss of sleep and loss of time for other tasks'. There is more to behavioural research (and to life!) than staying up all night watching a slug doing nothing! Shakespeare can have the last word on the subject. He might have been lamenting the plight of a poor research student, watching (not videoing) a reluctant subject at the behest of her/his respected but demanding supervisor:

Is it thy will thy image should keep open
my heavy eyelids to the weary night?
Dost thou desire my slumbers should be broken,
While shadows, like to thee, do mock my sight?

Sonnets LX1

Steve Wratten
Department of Entomology and Animal Ecology
Lincoln University
PO Box 84
Canterbury
New Zealand
July 1993

1

Flying insects in the field

J.R. Riley

1.1 INTRODUCTION

The problem of recording the daytime flight trajectories of insects in the field appears to have been first tackled by Sayer (1956) using multiple-exposure still photographs. Later, 16 mm cine cameras were used, with either normal frame rates (see references in Dahmen and Zeil, 1984), or with high-speed drives (Cooter and Baker, 1977; Baker *et al.*, 1984). Successful night-time observations were also made with cine cameras fitted with electronic image intensifiers and by using infra-red illumination (Murlis and Bettany, 1977; Murlis *et al.*, 1982), but both daylight and night-time cine methods required rather specialist expertise and equipment.

In recent years, the advent of cheap, easy to operate and portable video cameras and recorders has made the field study of insect flight a much more practicable proposition for the non-specialist. The technique offers a number of very significant practical advantages over cinematography, perhaps the most important being the facility to actually monitor camera image quality and to check recording integrity during observational experiments. Other advantages include the practicality of using automated procedures to track and digitize image position, sensitivity to very low illumination levels in both the visible and the near infra-red spectral regions, the cheap and reusable nature of the recording medium, and, in the case of stereoscopic recording, the relative ease with which synchronization of the frame rates of two cameras can be achieved. It thus seems very likely that video methods will be used by an increasingly large number of field entomologists interested in insect flight. The purpose of this chapter is to provide a convenient starting point for these users, and it is hoped that the material may also be of some use to scientists already familiar with video techniques. Emphasis

Video Techniques in Animal Ecology and Behaviour. Edited by Stephen D. Wratten. Published in 1993 by Chapman & Hall, London. ISBN 0 412 46640 6

will be placed on applications and techniques, rather than on video technology.

1.2 TYPES OF APPLICATION

Video recording has been used for two main categories of field observation: general descriptions of flight behaviour, usually in two dimensions, and quantitative flight trajectory measurements in three dimensions. Examples of the former include video studies of courtship behaviour in butterflies (Rutowski, 1978), moth flight in pheromone plumes (Kawasaki, 1981; David *et al.*, 1983), foraging performance (Seeley, 1983) and pattern recognition by honeybees (Lehrer *et al.*, 1985; Srinivasan and Lehrer, 1988), hovering flight by stingless bees (Zeil and Wittmann, 1989; Kelber and Zeil, 1990), swarming flight by mosquitoes (Peloquin and Olson, 1985), and estimation of the efficiency of light traps (McGeachie, 1988) and of suction traps (Schaefer *et al.*, 1985). Two-dimensional video recordings have also yielded quantitative descriptions of the behaviour of tsetse flies in the vicinity of odour plumes (Gibson and Brady, 1985, 1988; Brady *et al.*, 1990; Gibson *et al.*, 1991; Packer and Warnes, 1991), and the technique has been used to measure the efficiency of electric nets (Packer and Brady, 1990), and to investigate mating pursuit flights by tsetse flies (Brady, 1991).

Three-dimensional video measurements in the field have been reported much less frequently and, to date, appear to be limited to the work of Hollingsworth (1986), describing moth flight trajectories in the vicinity of pheromone plumes, and our studies (Riley *et al.*, 1990, 1992) dealing with both emigratory and 'vegetative' flight of moths over a host crop.

1.3 FACTORS AFFECTING THE VISIBILITY OF AIRBORNE INSECTS

Given the excellence of currently available video equipment, it is a straightforward matter to produce good-quality recordings of large, slow-moving, well-lit and high-contrast subjects. Unfortunately, insects do not usually fall into this category, and it is correspondingly rather more difficult to make successful records of their flight in the field.

1.3.1 Insect size and track length

The small size of insects does not in itself present an insurmountable problem because, with the appropriate choice of lens and viewing distance, even the smaller species can be resolved. However, size *does* affect the length of flight track observable. Thus, suppose that a lens is chosen that will produce the maximum field of view consistent with the formation of a 'spot' image large enough to be detected when the target insect is at a convenient range. In this condition, the image will register (at any one time) on not more than a single line of the 625 used in normal video systems. One would then expect that in the case of an ideal, flare-free system, the *maximum* length of straight-line flight trajectory observable in a plane perpendicular to the viewing axis of a stationary camera would be approximately 625 times an average body dimension in the vertical (screen) direction, and 1.3 times this distance across the screen. This figure is consistent with the experimental results of Gibson and Brady (1985), who found that they were able to record tsetse fly trajectories over an area of 2.5 × 2.5 m – dimensions some 600 times the 3–4 mm typical of fly body size.

Track length in the direction of the viewing axis will depend on the depth of focus of the camera lens. Viewing distance will usually be greater with larger insects and, as depth of focus increases with viewing distance, maximum track length in the viewing axis dimension will also tend to be related to body dimensions.

1.3.2 Background and contrast

It is important to note that Gibson and Brady's close-to-ideal performance was achieved only when the flies were seen, brightly lit, against the dark background presented by a black cloth. If insects are seen against a *light* background, reliable detection might require registration on up to 10 video scan lines (Hollingsworth, 1986), and the maximum trajectory would then be reduced to only 60–70 body dimensions. This is because the transient images of small targets moving against a light background tend to be 'filled in' by the after-image of the background. Contrast against a natural background may be increased if the target can be marked with a material of bright, contrasting colour. Thus, in their observations (from a 27 m tower) of *Lymantia dispar* following a pheromone plume, David *et al.* (1983) found that the visibility of the moths was increased when they were dusted with pink fluorescent powder.

Considerable advantage is to be gained by working with a brightly illuminated target against a dark background, because under these conditions the target image will register even if its size is well below the line-width of the video system, and the length of observable trajectory may thus be well above the maximum value quoted in section 1.3.1. The degree of increase is difficult to predict because it depends on the character of the camera photo-sensor and on the contrast between background and target, but we have found that factors of 2–3 are achievable when viewing insects illuminated against the background of the night sky. An additional advantage is that images of bright targets leave a trail of afterglow 'footprints' on the background, and this conveniently serves both to highlight movement and to indicate its direction (Fig. 1.1).

Figure 1.1 Stereoscopic video: a photograph of a split-screen video display of the views from two cameras separated by a distance of 35 m (see Fig. 1.2). The top half of the screen shows the view from the left-hand camera, and the bottom half that from the right. The diagonal lines are light fishing lines supporting calibration spheres (pingpong balls), and the stereo-pair images of a flying moth are the meteor-like traces near the centre of the screen. The afterglow of the bright image of the moth serves to highlight its movement (reproduced from Riley *et al.*, 1990, by permission of Blackwell Scientific Publications Ltd).

1.4 ARTIFICIAL ILLUMINATION

Although it is feasible to use ambient light to record video images of insects flying during the daylight, the combination of their small size and (frequently) low contrast against the natural background will often limit the length of observable flight trajectory to a few tens of body dimensions. In some circumstances it is possible to increase contrast and hence observable trajectory length by the introduction of an artificially dark background, either a simple black cloth (Gibson and Brady, 1985), or more elaborate (and more effective) non-reflective structures (Schaefer and Bent, 1984). In other circumstances – when viewing against the sky for example – the only way of increasing contrast is to use an artificial illumination source to increase the brightness of the target. Artificial illumination is, of course, usually required for nocturnal observations. Exceptionally, in the case of observations made with passive infra-red (PIR) viewing equipment, no illumination is required at all, because devices of this type form images from the (longer-wave) infra-red radiation *emitted* by bodies by virtue of their temperature. PIR equipment has in the past been rather cumbersome and very expensive, but the recent development of image-forming sensors that function at room temperature make it likely that PIR equipment will become much cheaper and more widely available in the future (Anon., 1992), and it will undoubtedly find application in entomological studies.

1.4.1 Wavelengths for illuminators

Many insects, especially night-flyers, respond strongly to bright lights, so it is obviously important that illuminators used for video studies operate in spectral regions which are outside the range of visual perception of the insect being studied. Laboratory experiments on the colour vision of a variety of insects show that their sensitivity is reduced at the red end of the visible spectrum, falling to very low values at wavelengths longer than 650 nm (Mikkola, 1972; Horridge *et al.*, 1975; Langer *et al.*, 1979; White, 1985). It has been suggested that some moths may be able to detect radiation in the far infra-red region (2000–20 000 nm), using their antennae as dielectric aerials (Laithwaite, 1960; Callahan, 1977), but this hypothesis is unsubstantiated (Kettlewell, 1961; Hsiao, 1972) and it is generally believed that, with perhaps a few exceptions (Evans, 1966; Bruce, 1971; Richerson and Borden, 1972), insects are sensitive to infra-red illumination only at power levels high enough to produce a heating effect. Infra-red has thus been accepted as the most appropriate wavelength range with which to illuminate insect targets during film and video studies. The sensitivity of image intensifier and camera photo-sensors falls to low values beyond 850–900 nm, so it is the shorter wavelength end of the near region of the infra-red spectrum (750–1500 nm) that is the wavelength range used in practice.

1.4.2 Illumination sources

All film and video studies of night-time flight reported to date have used incandescent tungsten filament lamps, equipped with filters which strongly absorb visible light but transmit wavelengths longer than about 700 nm. Filters found to be satisfactory include Kodak No 11 Safelight (Howell and Granovsky, 1982), Wratten 87 gelatine (Murlis and Bettany, 1977) and Schott RG 780 (Hollingsworth, 1986). Illuminator electric power consumption has ranged from 15 W (Howell and Granovsky, 1982) to 1000 W (Peloquin and Olson, 1985; Riley *et al.*, 1990). Weatherproofed illuminators, conveniently complete with suitable infra-red filters, can now be obtained from commercial sources active in the security/surveillance field.

Although simple filtered lamp illuminators have proved satisfactory for night-time observations, they become progressively less effective during dusk and dawn, as illumination from the natural background begins to reduce image contrast by competing with the illuminator light reflected from the target. The adverse effect of twilight illumination may be reduced somewhat by fitting a filter on to the video camera which is transparent to the infra-red emitted from the illuminator but which attenuates visible light. Unfortunately, natural illumination contains a substantial component of near infra-red radiation, so this strategy yields only a limited advantage. Full invulnerability to twilight requires the use of more specialized lighting equipment of the type employed by Schaefer and Bent (1984). Their source, which was able to illuminate small targets brightly enough to make them show up in high contrast against the midday sky, used a filtered xenon lamp which was pulsed at high power in synchronism with an electronic shutter in their intensifier/video-camera detector. By this means, average background illumination was reduced because light was admitted to the camera only during the brief (30 μs) periods when the illuminator was operating at peak power, and thus target contrast was greatly increased. Additional contrast was obtained by the use of a filter on the intensifier which discriminated further against background illumination, admitting wavelengths only within a narrow (2 nm) spectral band centred on an emission peak at 823 nm in the xenon flash output.

1.4.3 Spectral sensitivity of video cameras

Benefit from illumination in the near infra-red region will be gained only if the sensitivity of the video cameras being used extends into the spectral region above 700 nm. Fortunately, photo-sensors with sensitivity at wavelengths up to 1100 nm (Howell and Granovsky, 1982) have been developed, and cameras effective to 900 nm are readily available.

1.5 OBSERVATIONAL CONFIGURATIONS

The experimental configuration of cameras, and of background screens or illuminators (if used), will usually be dictated by the aspect of insect flight which it is intended to investigate, and by the constraints imposed by the visibility of the target species. For example, we wished to observe the nocturnal flight of *Helicoverpa armigera* moths as they emigrated from a host crop in which they had emerged (Riley *et al.*, 1992). Ideally, we would have liked to have observed the whole of the crop from an elevated position, and recorded the trajectories of all the emigrating moths. However, because the moths were not large and presented low contrast against the background of the crop, observations would have been practical over an area delineated by only a few tens of body lengths. Enhancement of contrast (and therefore of viewing area) by the use of infra-red illumination was possible, but only if the illuminator did not also light up the background. We thus adopted the compromise of viewing the moths brightly illuminated against the night sky by an infra-red searchlight (see Section 1.5.2 below), so that they formed high-contrast targets and could be reliably tracked over a volume of about 100 m³.

It is of relevance to note that the use of night-vision goggles has proved a valuable supplement to video equipment in nocturnal studies of moth behaviour (Lingren *et al.*, 1980).

1.5.1 Two-dimensional observations

There are many experiments where sufficient information about an insect's behaviour can be gained from a simple two-dimensional record of its flight trajectory. For instance, in McGeachie's (1988) investigation of light-trap efficiency, the behaviour of moths in the vicinity of the trap could be adequately described from the video image produced by a single camera pointing towards (but shielded from) the trap. Similarly, the departure and return of labelled honeybees from an experimental hive have been satisfactorily recorded with a single camera (Seeley, 1983), as has the flight of bees choosing between patterns in a hive arrival chamber (Srinivasan and Lehrer, 1988).

Quantitative descriptions of flight trajectories are possible with a single camera in cases where the insect flight path lies largely in one plane. Thus, Perry *et al.* (1988) were able to use a simple trigonometric method to reconstruct the horizontal flight path of moths over a substrate which was obliquely inclined to their camera's line of sight. Similarly, in the case of studies of moths (David *et al.*, 1983) and of tsetse flies (Gibson and Brady, 1985), detailed and quantitative descriptions of flight trajectories were possible using recordings from a single downward-pointing camera. Trajectory reconstruction requires extraction of the coordinates of the target within the video frame (see Section 1.5.2 below), and the use of a scaling factor deduced from calibration points or from the camera focal length and target range.

Knowledge of the local wind field is needed to interpret trajectories in terms of flight behaviour, and this was conveniently obtained by David *et al.* (1983) and by Gibson and Brady (1985) by including a grid of tethered windblown 'telltale' ribbons within the camera field of view. On other occasions, soap-bubbles and other airborne particles were used (David *et al.*, 1983; Brady *et al.*, 1990).

1.5.2 Three-dimensional observations

It is sometimes essential to record flight in three dimensions, and in these cases, some form of two-camera stereoscopic viewing system is normally required. The trigonometric basis on which such systems are based is shown in Fig. 1.2. The position of a remote target is defined by the distance

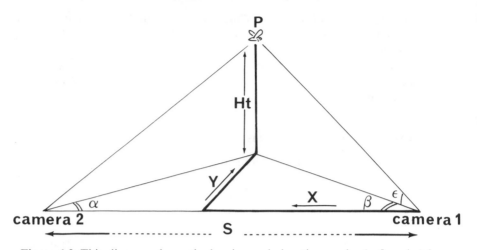

Figure 1.2 This diagram shows the bearing and elevation angles (α, β, and ε) from two camera positions, which fix the three-dimensional position coordinates (X, Y and Ht) of an airborne insect at P, relative to the right-hand camera.

$X = S \times \sin(\alpha) \times \cos(\beta)/\sin(\alpha + \beta)$

$Y = S \times \sin(\alpha) \times \sin(\beta)/\sin(\alpha + \beta)$

$Ht = S \times \sin(\alpha) \times \tan(\varepsilon)/\sin(\alpha + \beta)$

(reproduced from Riley *et al.*, 1990, by permission of Blackwell Scientific Ltd).

separating the two cameras and by the three angles (α, β and ε), and the function of the cameras is simply to record data from which these angles can be extracted. In the case of moving targets, the data from the two viewing positions must be recorded synchronously, and this may conveniently be done – albeit at the cost of half the field of view – by storing one half of the video field from each camera in a split-frame format (Hollingsworth, 1986; Riley *et al.*, 1990). Extraction of the required three angles can be achieved either by comparing the position of the target image in the two video pictures with the position of calibration points subtending known angles at the cameras (Riley *et al.*, 1990); by reference to the camera alignment angles and fields of view (Hollingsworth, 1986); or by the use of a three-dimensional calibration target (Dahmen and Zeil, 1984).

The target and calibration point positions can be measured from tracings from the screens or, more satisfactorily, by using a digitizer which produces movable cross-hairs on the screen and registers their coordinates. In either case, the first requirement is to identify those targets in the field of view from each video camera which form 'stereo-pair' images of individual insects (Fig. 1.1). Stereo pairs are usually easy to detect because they move in synchronism, turning together (although often in apparently different directions), and appearing and disappearing from the video fields of view at similar times. Once a stereo pair has been selected, two pairs of coordinates are logged (one from each video field), and the process is repeated for sequential frames;

the screen coordinates are then translated into angles and finally, using the equations in Fig. 1.2, into three-dimensional coordinates. Sequential sets of spatial coordinates define the insect trajectory, and its displacement speed can be obtained from trajectory length and the video frame rate.

Logging the image coordinates by repeatedly adjusting the position of cross-hairs is a time-consuming and rather tedious process, and can sometimes be avoided by the use of a target-following algorithm implemented on a small desktop computer (Schaefer and Bent, 1984), even in the presence of background images (J. Zeil, pers. comm.). A number of algorithms for target-following are available from commercial sources. The process of spatially calibrating the stereo viewing system can itself be carried out by computer, without the need for experimental calibration procedures, provided that five or more arbitrary but discrete points in space are simultaneously visible to both cameras (see comments in Dahmen and Zeil, 1984).

It is appropriate to mention here an ingenious *single-camera* technique for recording insect flight paths in three dimensions which was described by Okubo *et al.* (1981). The technique used a cine camera, but would work just as well with a video system, and took advantage of the trigonometric relation between the position of individual midges and the shadows they cast when flying above a white surface. It is also perhaps relevant to mention a simple alternative to stereoscopic video for recording the three-dimensional flight trajectories of butterflies. In this method, two observers keep the target insect in sight through two viewing tubes, whose elevation and bearing angles are recorded electronically. The three angles required for trajectory reconstruction are thus acquired with a minimum of equipment (Zalucki *et al.*, 1980).

1.6 TARGET IDENTIFICATION

Target identification presents problems in all remote sensing techniques (Riley, 1989), and the video recording of insect flight behaviour in the field is no exception. In some instances the problem is easily resolved, as in the case when observing labelled bees at close range (Seeley, 1983), but in others, when the observed insect appears simply as a featureless point target, positive identification from the video image alone may be difficult, if not impossible. In these situations, supplementary observations may sometimes be combined to identify the species most likely to be seen on the video at any one time. As a rule, the smaller the species of interest, the more difficult the problem of identification becomes, because of the greater range of species in the smaller size categories.

1.6.1 Image size

The size of the image on the video screen may provide a clue to the species of insect being detected, larger and more reflective insects tending to produce larger, brighter images. An indication of the size to be expected from the species under investigation may be gained by moving a specimen within the camera field of view at the range of interest. Image size depends, however, on the aspect presented by the insect and on its wing positions, and so at best is only a crude indicator of target size, and hence of species.

1.6.2 Wingbeat frequency

The action of wingbeating will modulate the light scattered from an insect, and the frequency of modulation may provide a clue to its species (Greenewalt, 1962; Schaefer, 1976; Corben, 1983; Oertli, 1991). The frame rate of 25 Hz normally used in video is too low to unambiguously sample the wingbeat frequency of any but the largest insects with wingbeats of less than 10 Hz. However, if an insect is moving and the reflection from its wings is large, the wingbeat action may register as a spatially distributed brightness modulation along the flight track, and its frequency can be deduced from the number of brightness cycles per video frame. In many cases, the amplitude of modulation and/or the speed of movement will be insufficient to produce this effect and, if wingbeat frequency is to be used as an aid to identification, it must be acquired using supplementary equipment. We found that the combination of an infra-red illuminator and a photomultiplier detector developed by Farmery (1982) (Fig. 1.3) functioned well as a wingbeat frequency monitor in our video and radar observations of moths (Rose *et al.*, 1985; Riley *et al.*, 1990, 1992), and the principle has been shown

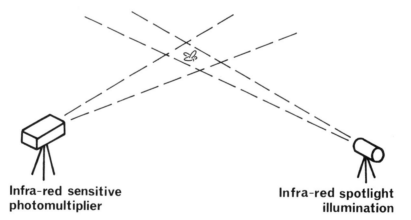

**Infra-red sensitive
photomultiplier**   **Infra-red spotlight
illumination**

Figure 1.3 Diagram showing the principle of an infra-red wingbeat frequency measuring device. The field of view of the photomultiplier (equipped with a telephoto lens) intercepts the infra-red beam from a focused illuminator. Any insect flying into the volume of interception scatters infra-red light, some of which is detected by the photomultiplier. The amplitude of the scattered light is modulated at the insect wingbeat frequency. The volume of interception must be chosen so that a reasonable rate of interception of single targets is achieved.

to work on insects as small as mosquitoes (Moore *et al.*, 1986). However, it should be emphasized that in environments where several species of similar size (and wingbeat frequency) may be airborne at the same time, frequency spread within species and overlap between them will severely limit the degree to which identification can be based on wingbeat frequency alone. This is especially the case if frequency is affected by changes in air temperature (Farmery, 1982; Unwin and Corbet, 1984; Oertli, 1989) or by changes in flight behaviour such as pheromone-following (Riley *et al.*, 1992).

1.6.3 Characteristic behaviour

In a few experimental situations, the identity of the insects recorded on video is made clear by their behaviour. This is especially the case in studies of moths approaching pheromone sources, because only males of the target species can be expected to show any reaction to the source (Lingren *et al.*, 1980; Murlis *et al.*, 1982; Hollingsworth, 1986). Similarly, the response of tsetse flies to the release of host odour has been found to provide an unambiguous clue to their identity (Gibson and Brady, 1985). The swarming behaviour of mosquitoes has also proved to be a convenient identification feature (Peloquin and Olson, 1985). In some cases it may be possible to closely approach insects which are seen to settle, and to identify them by eye (or with the aid of night-vision binoculars).

1.6.4 Supplementary trapping

We found it useful in our experiments to sample the species airborne at, or just below, the altitude being observed by video, by using a net mounted above a motor vehicle (Drake, 1990; Riley *et al.*, 1992). The netting was carried out round the periphery of the experimental area, and so the samples could not be guaranteed to exactly represent the aerial fauna over the centre, where the video system was operating. Fortunately, however, the speed and generally linear nature of the observed flight trajectories made it unlikely that there would be any significant gradient of aerial population over the dimensions of the experimental area, and so we were able to obtain a good estimate of the species composition of the population recorded on the video cameras.

Aerial trapping by model aircraft (Tedders and Gottwald, 1986) may perhaps be appropriate if high densities of small insects are being studied, and in the case of larger insects, large-area stationary 'goal-net' traps have proved useful (Drake, 1990).

1.7 SUMMARY

The major limitations of video as a method of observing insect flight are its short effective range (usually less than 10 m) and its inability to resolve high-speed movements (Rutowski, 1978). In studies where these limitations are not important, the technique provides a convenient and versatile means of recording and analysing flight behaviour in the field, and it has been successfully used to observe insects ranging in size from aphids to butterflies.

Advances in video technology continually produce improvements in performance and reductions in both the cost and size of equipment, so it seems certain that the technique will prove to be an even more useful resource to the field entomologist in the future.

REFERENCES

Anon. (1992). A room temperature infra-red camera. *Electronic Engineering*, **64**(781), 9–10.

Baker, P.S., Gewecke, M. and Cooter R.J. (1984). Flight orientation of swarming *Locusta migratoria*. *Physiological Entomology*, **9**, 247–252.

Brady, J. (1991). Flying mate detection and chasing by tsetse flies (*Glossina*). *Physiological Entomology*, **16**, 153–161.

Brady, J., Packer, M.J. and Gibson, G. (1990). Odour plume shape and host finding by tsetse. *Insect Science and its Application*, **11**(3), 377–384.

Bruce, W.A. (1971). Perception of infrared radiation by the spiny rat mite *Laelaps echidnina* (Acari: Laelapidae). *Annals of the Entomological Society of America*, **64**(4), 925–931.

Callahan, P.S. (1977). Moth and candle: the candle flame as a sexual mimic of the coded infrared wavelengths from a moth sex scent (pheromone). *Applied Optics*, **16**(12), 3089–3097.

Cooter, R.J. and Baker, P.S. (1977). Weis–Fogh clap and fling mechanism in *Locusta*. *Nature*, **269**(5623), 53–54.

Corben, C.C. (1983). Wingbeat frequencies, wing areas and masses of flying insects and hummingbirds. *Journal of Theoretical Biology*, **102**, 611–623.

Dahmen, H.-J. and Zeil, J. (1984). Recording and reconstructing three-dimensional trajectories: a versatile method for the field biologist. *Proceedings of the Royal Society of London B*, **222**, 107–113.

David, C.T., Kennedy, J.S. and Ludlow, A.R. (1983). Finding of a sex pheromone source by gypsy moths released in the field. *Nature*, **303**, 804–806.

Drake, V.A. (1990). Methods for studying adult movement in *Heliothis*, in Heliothis: *Research Methods and Prospects*, (ed M.P. Zalucki), Springer-Verlag, London, pp. 109–121.

Evans, W.G. (1966). Perception of infrared radiation from forest fires by *Melanophila acuminata* De Greer (Buprestidae, Coleoptera). *Ecology*, **47**, 1061–1065.

Farmery, M.J. (1982). The effect of air temperature on the wingbeat frequency of naturally flying armyworm moths (*Spodoptera exempta*). *Entomologia Experimentalis et Applicata*, **32**, 193–194.

Gibson, G. and Brady, J. (1985). 'Anemotactic' flight paths of tsetse flies in relation to host odour: a preliminary video study in nature of the response to loss of odour. *Physiological Entomology*, **10**, 395–406.

Gibson, G. and Brady, J. (1988). Flight behaviour of tsetse flies in host odour plumes: the initial response to leaving or entering odour. *Physiological Entomology*, **13**, 29–42.

Gibson, G., Packer, M.J., Steullet, P. and Brady, J. (1991). Orientation of tsetse flies to wind, within and outside host odour plumes in the field. *Physiological Entomology*, **16**, 47–56.

Greenewalt, C.H. (1962). Dimensional relationships for flying animals. *Smithsonian Miscellaneous Collections*, **144**(2).

Hollingsworth, T.S. (1986). The influence of local wind effects upon the approach behaviour of some male Lepidoptera to field pheromone sources. PhD thesis, Cranfield Institute of Technology, Bedfordshire, England.

Horridge, G.A., Mimura, K. and Tsukhara, Y. (1975). Fly photoreceptors. II Spectral and polarised light sensitivity in the drone fly *Eristalis tenax*. *Proceedings of the Royal Society B*, **190**, 225–237.

Howell, H.N. and Granovsky, T.A. (1982). An infrared viewing system for studying nocturnal insect behaviour. *The Southwestern Entomologist*, **7**(1), 36–38.

Hsiao, H.S. (1972). The attraction of moths (*Trichoplusia ni*) to infrared radiation. *Journal of Insect Physiology*, **18**, 1705–1714.

Kawasaki, K. (1981). A functional difference of the individual components of *Spodoptera litura* (F.) (Lepidoptera: Noctuidae) sex pheromone in the attraction of flying male moths. *Applied Entomology and Zoology*, **16**(2), 63–70.

Kelber, A. and Zeil, J. (1990). A robust procedure for visual stabilisation of hovering flight position in guard bees of *Trigona* (*Tetragonisca*) *angustula* (Apidae, Meliponinae). *Journal of Comparative Physiology A*, **167**, 569–577.

Kettlewell, H.B.D. (1961). The radiation theory of female assembling in the Lepidoptera. *The Entomologist*, **94**, 59–65.

Laithwaite, E.R. (1960). A radiation theory of the assembling of moths. *The Entomologist*, **93**, 113–117, 133–137.

Langer, H., Hamann, B. and Meinecke, C.C. (1979). Tetrachromatic visual system in the moth *Spodoptera exempta* (Insecta: Noctuidae). *Journal of Comparative Physiology*, **129**, 235–239.

Lehrer, M., Wehner, R. and Srinivasan, M. (1985). Visual scanning behaviour in honey bees. *Journal of Comparative Physiology A*, **157**, 405–415.

Lingren, P.D., Burton, J., Shelton, W. and Raulston, J.R. (1980). Night vision goggles: for design, evaluation and comparative efficiency determination of a pheromone trap for capturing live adult male pink bollworms. *Journal of Economic Entomology*, **73**(5), 622–630.

McGeachie, W.J. (1988). A remote sensing method for the estimation of light-trap efficiency. *Bulletin of Entomological Research*, **78**, 379–385.

Mikkola, K. (1972). Behavioural and electrophysiological responses of night-flying insects, especially Lepidoptera, to near ultra-violet and visible light. *Annals Zoologica Fennici*, **9**, 225–254.

Moore, A., Miller, J.R., Tabashnik, B.E. and Gage, S.H. (1986). Automated identification of flying insects by analysis of wingbeat frequencies. *Journal of Economic Entomology*, **79**, 1703–1706.

Murlis, J. and Bettany, B.W. (1977). Night flight towards a sex pheromone source by male *Spodoptera littoralis* (Boisd.) (Lepidoptera, Noctuidae). *Nature*, **268**, 433–434.

Murlis, J., Bettany, B.W., Kelly, J. and Martin, L. (1982). The analysis of flight paths of male Egyptian cotton leafworm moths, *Spodoptera littoralis*, to a sex pheromone source in the field. *Physiological Entomology*, **7**, 435–441.

Oertli, J.J. (1989). Relationship of wing beat frequency and temperature during take-off flight in temperate-zone beetles. *Journal of Experimental Biology*, **145**, 321–338.

Oertli, J.J. (1991). Interspecific scaling (relative size change) of wing beat frequency and morphometrics in flying beetles (Coleoptera). *Mitteilungen der Schweizerischen Entomologischen Gesellschaft*, **64**, 139–154.

Okubo, A., Bray, D.J. and Chiang, H.C. (1981). Use of shadows for studying the three-dimensional structure of insect swarms. *Annals of the Entomological Society of America*, **74**, 48–50.

Packer, M.J. and Brady, J. (1990). Efficiency of electric nets as sampling devices for tsetse flies (Diptera: Glossinidae). *Bulletin of Entomological Research*, **80**, 43–47.

Packer, M.J. and Warnes, M.L. (1991). Responses of tsetse to ox sebum: a video study in the field. *Medical and Veterinary Entomology*, **5**, 23–27.

Peloquin, J.J. and Olson, J.K. (1985). Observations on male swarms of *Psorophora columbiae* in Texas ricelands. *Journal of the American Mosquito Control Association*, **1**(4), 482–488.

Perry, J.N., Wall, C. and Clark, S.J. (1988). Close-range behaviour of male pea moths, *Cydia nigricana*, responding to sex pheromone re-released via the substrate. *Entomologia Experimentalis et Applicata*, **49**, 37–42.

Richerson, J.V. and Borden, J.H. (1972). Host finding by heat perception in *Coeloides brunneri* (Hymenoptera: Braconidae). Canadian Entomologist, **104**, 1877–1881.

Riley, J.R. (1989). Remote sensing in entomology. *Annual Review of Entomology*, **34**, 247–271.

Riley, J.R., Smith, A.D. and Bettany, B.W. (1990). The use of video equipment to record in three dimensions the flight trajectories of *Heliothis armigera* and other moths at night. *Physiological Entomology*, **15**, 73–80.

Riley, J.R., Armes, N.J., Reynolds, D.R. and Smith, A.D. (1992). Nocturnal observations of the emergence and flight behaviour of *Helicoverpa armigera* (Lepidoptera: Noctuidae) in the post-rainy season in central India. *Bulletin of Entomological Research*, **82**, 243–256.

Rose, D.J.W., Page, W.W., Dewhurst, C.F. *et al.* (1985). Downwind migration of the African armyworm moth, *Spodoptera exempta*, studied by mark-and-capture and by radar. *Ecological Entomology*, **10**, 299–313.

Rutowski, R.L. (1978). The courtship behaviour of the small sulphur butterfly, *Eurema lisa* (Lepidoptera: Pieridae). *Animal Behaviour*, **26**, 892–903.

Sayer, H.J. (1956). A photographic method for the study of insect migration. *Nature, London*, **177**, 226.

Schaefer, G.W. (1976). Radar observations of insect flight, in *Insect Flight*, (ed. R.C. Rainey), Symposium of the Royal Entomological Society of London, No. 7, Blackwell Scientific Publications, Oxford, pp. 157–197.

Schaefer, G.W. and Bent, G.A. (1984). An infra-red remote sensing system for the active detection and automatic determination of insect flight trajectories (IRADIT). *Bulletin of Entomological Research*, **74**, 261–278.

Schaefer, G.W., Bent G.A. and Allsop, K. (1985). Radar and opto-electronic measurements of the effectiveness of the Rothamstead Insect Survey suction traps. *Bulletin of Entomological Research*, **75**, 701–715.

Seeley, T.D. (1983). Division of labor between scouts and recruits in honeybee foraging. *Behavioural Ecology and Sociobiology*, **12**, 253–259.

Srinivasan, M.V. and Lehrer, M. (1988). Spatial acuity of honeybee vision and its spectral properties. *Journal of Comparative Physiology*, **162**, 159–172.

Tedders, W. L. and Gottwald, T.R. (1986). Evaluation of an insect collecting system and an ultra-low volume spray system on a remotely piloted vehicle. *Journal of Economic Entomology*, **79**(3), 709–713.

Unwin, D.M. and Corbet, S.A. (1984). Wingbeat frequency, temperature and body size in bees and flies. *Physiological Entomology*, **9**, 115–121.

White, R.H. (1985). Insect visual pigments and color vision, in *Comprehensive Insect Physiology, Biochemistry and Pharmacology*, Vol. 6, *Nervous System: Sensory*. (eds G.A. Kerkut and L.I. Gilbert), Pergamon, Oxford, pp. 431–493.

Zalucki, M.P., Kitching, R.L., Abel, D. and Pearson, J. (1980). A novel device for tracking butterflies in the field. *Annals of the Entomological Society of America*, **73**, 262–265.

Zeil, J. and Wittman, D. (1989). Visually controlled station-keeping by hovering guard bees of *Trigona* (*Tetragonisca*) *angustula* (Apidae, Meliponinae). *Journal of Comparative Physiology A*, **165**, 711–718.

2

Flying insects in the laboratory

S. Young, J. Hardie and G. Gibson

This chapter examines the methods used to make videos of insects flying in large cages ('arenas') and in the working sections of wind tunnels. It first addresses some general principles of designing a video setup, the acquisition of data from video tapes and then reports some illustrative case studies.

2.1 EQUIPMENT

Solid state (CCD) cameras are best. They are small, light, very sensitive in the infra-red region and need only low-voltage power supplies. Always choose a camera which is capable of accepting external synchronizing pulses (see Fig. 2.1), so that, if you do end up with two cameras (see below), it will be possible to keep them exactly in step with each other. A laboratory camera does **not** need:

- to be colour (black and white has better resolution; colour contrast can be achieved by putting a suitable colour filter on to a black and white camera);

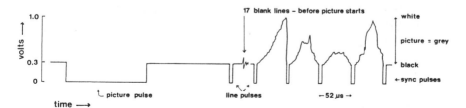

Figure 2.1 A video signal has two regions: voltages below 0.3 V are synchronizing pulses which tell the monitor when to start a new picture or line, and voltages above 0.3 V code a grey scale. Because of the blank lines during the picture pulse and before the picture begins, there are really only 575 scan lines in a 625-line picture.

Video Techniques in Animal Ecology and Behaviour. Edited by Stephen D. Wratten. Published in 1993 by Chapman & Hall, London. ISBN 0 412 46640 6

- to have a viewfinder (use the monitor);
- to have an auto-iris (self-adjusting) lens (the light is controlled and optimum settings for seeing small bright dots are generally not what the servo would select).

The rest of the setup comprises a time/date marker box (which super-imposes the time and date on to the picture), a video recorder and a TV monitor. Because video signals are quite small but very fast-changing voltages (Fig. 2.1), it is not easy to send them from place to place. Coaxial cable reduces interference but has considerable capacitance and tends to attenuate fast-changing signals. The solution is to terminate the cable – short a 75 Ω resistor across it – at every input socket, and to provide every output socket with an amplifier powerful enough to drive this load. All commercial video equipment incorporates these devices. In consequence, you must always connect video equipment in a daisy chain, with 'video out' sockets leading to 'video in' sockets. Typically, the camera would link to the marker box, which would link to the recorder, which would link to the monitor. Time/date marking is essential for successfully retrieving informa-tion from video tapes. Ideally, the clock should show hundredths of a second, so that each individual picture has a separate label and irregularities in the video recorder's advance mechanism in the pause mode can be detected. Some video recorders incorporate a time/date marker facility which is generally only accurate to the nearest second, but this is often adequate.

The main requirement for the video recorder is to produce a high-quality picture in pause (single-frame) mode, and to have good controls for scanning forwards and backwards along the video tape, one picture at a time. Most recorders have a control for minimizing judder on single frame pictures (often on the circuit board inside the box), which may need adjusting to suit the monitor, and especially for loading pictures into a computer frame store. Video cameras actually produce pictures with 312.5 scan lines 50 times a second, and adjacent pairs of these 'half fields' interlock, giving a 625-line picture (Fig. 2.2). Some recorders in pause mode will only play back 25 pictures per second (generally they sample one half field twice, then miss out the next), which is a marked disadvantage with fast-flying insects.

Better-quality VHS domestic machines are just about usable for research (they must have genuine video inputs and outputs, not antenna sockets), but SVHS machines are markedly superior. U-matic is the professional stan-dard, which uses much larger (more expensive) cassettes.

Under certain circumstances two cameras are valuable: for example, situations when it is convenient to record instrument readings (e.g. wind speed and direction) on the video as well as pictures of insects and for 3D tracks (see also Chapter 1). When two cameras are used the signals can be fed into separate video recorders, as long as the cameras share the same sync

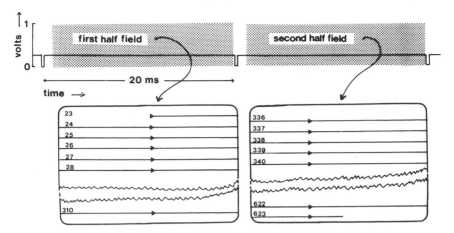

Figure 2.2 Successive half fields interlace to give a finer scanning grid, so that a complete picture (although with halved vertical resolution) is produced 50 times a second.

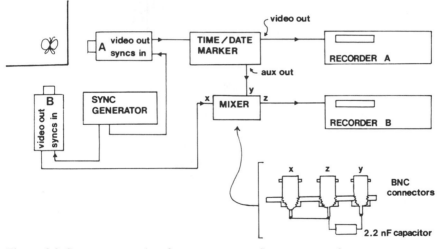

Figure 2.3 Separate recorders for two cameras. Some cameras have a sync output socket and a second camera can be 'slaved' to them by connecting via the sync output/input and eliminating the sync generator. The time/date marker box must have an auxiliary output socket giving a signal with just the time and date on a black background, for mixing with the video signal from camera B. It is then possible to use a simple mixer, just injecting the time/date signal via a small capacitor.

pulses and the same time/date marker trace is superimposed on both pictures (Fig. 2.3). Indeed, this is the only way to avoid some loss of resolution. It is, however, simpler to connect both cameras to a mixer box, split the picture into two zones ('wipe'), and feed each camera's signal to a separate zone (Fig. 2.4). It is sometimes possible to manage with a single

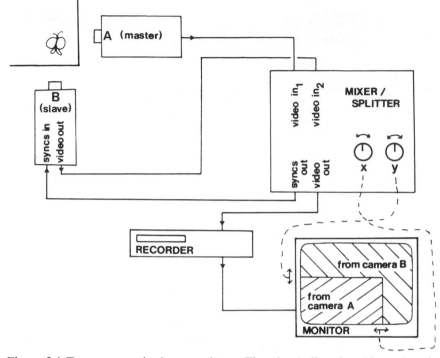

Figure 2.4 Two cameras sharing one picture. The mixer/splitter box (shown here in splitter mode) separates the sync pulses from camera A's signal and uses them to slave camera B. Knobs on the box control how the picture is divided up between the two cameras. Simple screen splitters are not capable of electronically compressing the whole picture from camera A, say, into a small fraction of the combined picture. In this case, they simply paste in the appropriate segment of the original picture and discard the rest.

camera and mirrors for 3D, providing the light path lengths and thus fields of view are of equal dimensions (see below).

2.2 OPTIMIZING IMAGES

The proper arrangement of the camera, lighting and background are critically important and can mean the difference between a clear, sharp picture and a hazy one in which the subject of interest seems to periodically disappear into the background. The main guidelines which should help to achieve a useful video recording and which will be discussed in this section are:

1. Provide adequate lighting for the insect and for the camera (the requirements for each may not necessarily be the same!).
2. Arrange lighting for the insects so as to minimize disorientation caused by light in the 'wrong' part of the insect's field of view.

3. Provide the insect with enough visual cues to allow natural behaviour – choose a field of view which is large enough to follow the insect wherever it goes, but small enough so that the insect does not disappear between the lines of the monitor.
4. Minimize reflections which would otherwise obscure the image of the moving insect in some areas of the picture.
5. Maximize contrast between the insect and the background.

2.2.1 Lighting

The need to provide insects with suitable visual stimuli frequently conflicts with a lighting regime which will give a good quality picture. It may therefore be necessary to consider the lighting requirements for the camera and for the insects separately: the position of the light source, the light intensity and the wavelength of light required for each can be quite different. Fortunately, the new solid-state cameras are very sensitive to long-wavelength light (i.e. above 900 nm) which is outside the spectral sensitivity range of insects. This means that infra-red light can be used to produce the optimal image quality on video, even when no light stimulus is visible to the insects. Chapter 3 gives examples of this for predatory insects.

For the camera, a ring of infra-red light-emitting diodes (IR LEDs) placed around the lens, or strips of IR LEDs placed inside the flight arena, can provide enough illumination on their own, especially if the light beams are arranged so that as much light as possible reflects off the insect's body into the camera. A ring of 20 high-powered IR LEDs (similar to those used in burglar alarms) is sufficient to illuminate an area of about 2 m^2 brightly enough to observe mosquitoes in the dark.

Unless the camera can be sited inside the arena, reflection may also pose problems. Many laboratory videos are shot through a window, which will reflect back any light coming from a source close to the camera. If such an arrangement is necessary, the reflection can be reduced by using a curved surface opposite the camera, or by aiming the camera at a steep angle with respect to the window.

Infra-red lights can also be used to backlight the flight arena, so that the flying insect appears as a silhouette against a light background. Backlighting is, however, generally less preferable, partly because it is more difficult to detect small dark objects on video than light ones (see Chapter 1) and also because far more infra-red light is needed to illuminate the entire field of view adequately. This technique was used by Gibson (1985), however, to record the behaviour of individual mosquitoes in a swarm over a stationary marker. A bank of $16 \times 15 \text{ W}$ incandescent strip lights covered with 'far-red' filters (passing wavelengths $> 680 \text{ nm}$, Ilford 609) was adequate to illuminate mosquitoes flying in a 1 m^3 cage.

Infra-red illumination for the camera is essential when studying the behaviour of night- or twilight-flying insects because appropriate ambient light levels will be too low for the camera. Even with day-flying insects in bright ambient light, a useful strategy is to use infra-red lights and to cover the camera's lens with an infra-red pass filter to remove visible wavelengths. Imaging is now purely a function of infra-red illumination, and the visible light fields can be optimized for the test insects without compromising picture quality (see Chapter 1).

Insects generally require diffuse lighting from above: point sources of light can be attractive in themselves and light which comes from the sides or underneath can be disorientating. The ideal light source would be an evenly illuminated opaque dome, something like the inside of a ping pong ball illuminated by the sky. A similar effect can be obtained by covering the flight arena with opal Perspex or white material, with incandescent lightbulbs or strip lights some distance away. Fluorescent lights are not recommended, because they have an on–off flicker at 100 Hz, i.e. at a slower rate than the flicker fusion frequency of most insect eyes (Miall, 1978).

2.2.2 Camera position and lens choice

Four factors influence the positioning of the camera and the choice of lens:

1. Covering the whole of the relevant area;
2. Avoiding background and foreground clutter, which will limit the zones in which the insects are clearly visible;
3. Minimizing distortions due to exaggerated perspective angles;
4. The physical dimensions of the laboratory.

Attempting to frame pictures properly with an actual camera is not at all easy and it is helpful to draw scale plans and elevations of the setup. First measure the angular size of the field of view of the lens in the horizontal and vertical planes by placing test cards at various distances from the camera (Fig. 2.5A). The field of view can then be matched to scaled drawings of the flight chamber (Fig. 2.5B). If the far wall of the chamber is chosen to fill the monitor screen, there will be blind areas towards the front of the chamber (Fig. 2.5B). If the camera is placed far enough away to view the whole of the chamber, the image will include a perspective view of the side walls, floor and ceiling (Fig. 2.5C).

Figure 2.5 Modelling the camera field of view and video setup. (A) Measure the angular size of the field of view in horizontal and vertical planes by placing test pattern cards at various distances from the lens. This is more accurate than relying on lens size, since the usable dimensions of the picture are limited by the monitor used. (B) The field of view just covers the back wall of the chamber; insects will only be seen in a portion of the chamber (stippled), although the

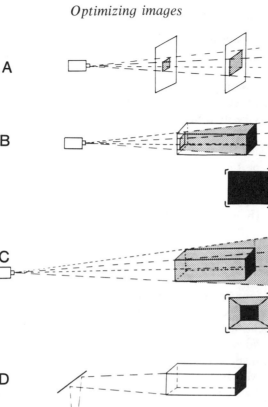

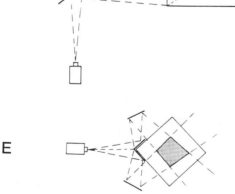

Figure 2.5 *contd.*

picture on the monitor will show that the back wall may provide an ideal even background (monitor picture in brackets). (C) The field of view includes entire chamber; the picture on the monitor will include areas of the chamber which may not prove an adequate background. (D) A mirror can be used to reposition the camera. Fold the cut-out model of the camera's field of view in the horizontal and vertical planes separately; the fold indicates the dimensions and position of the mirror. The field of view is the same as (D) above. (E) Mirrors can also be used to observe an area in 3D with only one camera. Again, use models to determine the dimensions and positions of the mirrors. The resulting space viewed will be rhomboid in cross-section (stippled).

The ideal is to use a matt black background in deep shadow with the insect illuminated perpendicularly to the line of view and a long telephoto lens on a camera placed as far as possible from the chamber. Insects will appear as bright objects on a dark background and will not change their apparent size as they move towards and away from the camera. Laboratory dimensions may be limiting: it is often not possible to place the camera far enough away, especially if viewing down from the ceiling. It is possible, however, to 'bend' the camera's field of view with mirrors (Fig. 2.5D) to fit the shape of the laboratory, but more than one reflection reduces image contrast. To determine the size and position of the mirror(s) required, fold a paper cutout of the field of view to fit the dimensions of the laboratory (Figs. 2.5C and D). The fold(s) are equivalent to the mirror's position.

2.2.3 Background

Achieving a homogeneous background is often impossible, as insects may well require patterned stimuli to maintain flight. Depending upon the plane in which the behaviour to be studied takes place, this patterned surface may end up as the videoed background. In such cases it is necessary to limit the contrast of the display as seen by the camera, while maximizing the stimulus the insects receive. One option is to use two colours for the patterned area of the background which contrast with each other from the point of view of the insect, when illuminated with white light, but which are of lower contrast for the camera, e.g. yellow and red.

All colours appear to be grey to the camera, and hence the contrast with the insect image will be greatest if the background is black or white. This can pose problems for automatic teletrackers, which require that the subject to be tracked is always either lighter or darker than the background. Unless the two-colour patterned background is very evenly illuminated, the insect may become of opposite contrast to the background in particular areas of the picture, which causes problems if image analysis or automatic tracking is to be used (see Section 2.3.1). A more satisfactory solution, although it requires manual analysis, is to make the background pattern black and white and to illuminate the camera's field of view with enough light so that the insect appears to be darker than the white areas and lighter than the dark areas. This is perhaps the only way to ensure that the insect receives the optimal visual inputs.

It is also possible that the patterned surface may end up as the window through which the camera should view. In this case, the insects may be viewed through a visual pattern on the floor or side. The pattern can be made with pieces of cellulose acetate or scrim (thin black plastic with fine holes, both available from stage-lighting companies) on the clear Perspex walls or floor of the flight arena. The camera will see the insects through either of these and, as long as there is a curtain of white material behind the camera, the pattern will be visible to the insects inside the flight arena (see Section 2.4.2 and Fig. 2.8). Orange cellulose acetate is relatively translucent to a camera which is illuminated with infra-red light, and scrim of various degrees of transparency can be obtained.

2.3 ANALYSING VIDEOS

2.3.1 Digitizing

The quantification process begins with digitization – producing a list of coordinates for the position in space of the insect at successive equal time intervals. The most primitive method of doing this is to use the pause mode of the video recorder, stick a piece of clear acetate sheet to the monitor screen, and ring your insect with a marker pen. Advance five pictures (say), and do it again, numbering every tenth sample. When the track becomes superimposed, a new acetate sheet is required. For this to be reasonably accurate, a reference frame of known actual dimensions should be traced on to each acetate sheet. A computer bit pad can then be used to input successive insect positions. It helps to get the computer to reconstruct the track as data points are added. However, this process is painfully slow – a minute of video can easily take a couple of hours to digitize – and as this is the main limiting factor in video research, it is worth trying to speed it up. The first step is to eliminate the acetate sheets and input coordinates directly to the computer. A 'pointer' is lined up with the insect image on a single frame, and a button is pressed to send the X and Y positions to the computer. It is best if the pointer is superimposed as a cross-hair or dot on the video signal (by a digitizer box connected between the recorder and the monitor) and steered with a joystick or mouse, but the simple arrangement shown in Fig. 2.6 works well.

The ideal is a mechanism which enables the computer to do the tracking itself, preferably at normal replay speed, and two techniques are available: teletracking and image analysis.

The teletracker works on standard video signals. A brightness threshold is set and an electronic comparator compares the incoming video signal with this threshold. It also records how far the scan has progressed when this level is first exceeded in a given picture, producing X and Y coordinates for the first bright dot encountered. This method works very well with pictures of a single small bright object on a black ground (see Section 2.4.3), but gives endless trouble with non-uniform backgrounds or groups of insects.

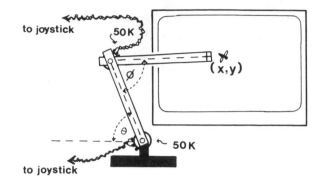

Figure 2.6 A low-cost digitizer. Use a games card and a joystick with your personal computer. Wire in the two potentiometers used to pivot the arms in place of those supplied with the joystick. The resistances logged by the computer are proportional to the angles θ and ϕ. $x = \cos(\phi - \theta) - \cos(\theta)$; $y = \sin(\theta) + \sin(\phi - \theta)$. Use the button to trigger a reading – it helps to average ten or so successive conversions of the reading from each potentiometer.

Image analysis equipment allows a frame-grabber to load each successive picture into the computer memory, representing it by a 512×512 array of numbers, each representing how bright a given spot or 'pixel' is. Software exists to process such digital images – for instance, it is possible to subtract the last picture's array from the current one, leaving only the moving objects, or to transform the image and enhance objects with sharp boundaries and lose out-of-focus background. Ultimately the picture is binarized (reduced to an array of 0s (black) and 1s (white)) and individual objects are automatically located and measured. Getting all this to happen in 20 ms is still something of a challenge – our image analyser makes a good job of tracking walking insects, when two samples a second is adequate. The main advantage of this method is that a sensible strategy can be implemented for when it loses the object it is tracking, and rules decided for difficult cases, for example when two tracks cross.

2.3.2 Extracting track parameters

Having acquired a collection of digitized tracks, it is necessary to compute the headings, speeds, turning rates and possibly some measure of the ground covered (area searched) in a given time. Some of the measures we use, with methods for calculating them, are shown in Fig. 2.7.

2.4 SPECIALIZED FLIGHT CHAMBERS

The simplest case is when the insects are free-flying in a large cage and the camera is placed inside or outside the arena. Such arrangements allow

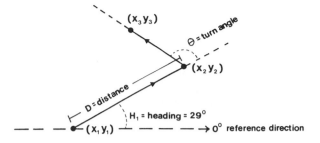

Figure 2.7 Calculating track parameters.

$D = \sqrt{(X_1 - X_2)^2 + (Y_1 - Y_2)^2}$

$H = \arcsine (X_2 - X_1/D)$. If $H < 0$, $H = H + 180°$. If $(Y_2 - Y_1) < 0$, $H = H - 360°$

$\theta = H_2 - H_1$ (negative for right turns, positive for left turns)

If $\theta > 180°$, $\theta = 360 - \theta$

If $\theta < -180°$, $\theta = (-360) - \theta$

For a sample lasting t seconds:

Speed $= \Sigma D/t$ cm/s

Turning rate $= \Sigma|\theta|/t°$/s

Tortuosity $= \Sigma|\theta|/\Sigma D°$/cm

Circling tendency $= \Sigma \theta/t°$/s

Ground covered $= \pi.(x$ interquartile).(y interquartile)

take-off behaviour (Grace and Shipp, 1988) and flight behaviour to be examined in still air.

Case histories of three specialized flight chambers are described. They were designed to answer specific experimental questions, but exemplify the basic principles laid out above.

2.4.1 Moving-floor wind tunnels

Moving-floor wind tunnels have been used to investigate the response of insects to odour plumes (Kennedy and Marsh, 1974; Kennedy *et al.*, 1981). Flying insects often use visually guided anemotaxis (upwind movement) to locate the source of an attractive odour. The insect controls its speed and direction of flight by responding to the apparent movement of the visual world around them (optomotor responses). Experiments in these horizontal wind tunnels often utilize a powerful odour source to attract the insect upwind. Thus, a male moth flying in a sex pheromone plume can be 'held' within the camera's field of view by controlling the speed of movement of a patterned floor beneath it. In this way the moth's behaviour can be observed over longer flight periods.

In such experiments the camera is placed above the tunnel and strip lights are used to illuminate both the camera's and the insect's field of view. These are placed inside the tunnel, along the upper edges, outside the camera's field of view. The floor of the tunnel is a conveyor belt, painted yellow with red circles; these colours are of high contrast to moth eyes, but appear to be similar shades of grey to a black and white video camera. Circles are used to avoid giving the moths any directional bias. The speed and direction of the floor is manually controlled so as to keep the insect in the field of view of the camera. Such is the strength of the visually guided anemotactic response that when insects enter an attractive odour plume they may even fly downwind and away from the odour source if the floor is moved in the upwind direction, against the actual air flow. This phenomenon has been shown, for example, in moths in sex pheromone plumes (Kennedy and Marsh, 1974) and tsetse flies in host odour plumes (Colvin *et al.*, 1989).

Moving floors are expensive and technically difficult to build. A similar effect can be created by projecting movement on to a stationary floor. A length of 35 mm film with clear and coloured patches is passed across a slide projector so as to cast a pattern of light and dark areas across the translucent floor of a wind tunnel from beneath. In this case, the image of the flying insect will appear through the camera as a silhouette against the illuminated floor, unless the light coming from above is very much brighter than the projector light.

2.4.2 'Barber's pole' tunnel

The 'barber's pole' tunnel was designed to test the optomotor response of free-flying insects (Fig. 2.8; David, 1982), based on the same principle as the moving-floor tunnel. Thus, even in still air, insects fly at a constant rate over the ground by adjusting their motor output in response to the apparent movement of the ground beneath them, so as to sustain a constant ground speed. If the ground speed is mechanically altered by moving visual patterns past a flying insect, the insect responds as though wind has blown it off course and alters its flight speed and direction to compensate. For example, consider insects flying in a horizontal, clear Perspex cylinder, in still air, within a larger tube with a striped, helical barber's pole painted on it (Fig. 2.8). If the outer tube is rotated, so as to create the impression that a series of stripes is moving from right to left, the insects alter their flight speed and direction. They fly either left to right slowly or right to left much faster, until they attain the preferred flight speed in relation to the stripes around them.

Although this apparatus was first used to investigate details of the optomotor response, it is also valuable for testing whether or not the light source and visual cues used in a flight arena are adequate for the insect to orientate. If the light intensity in the barber's pole tunnel is too low, the wavelengths of light used are outside of the visual spectrum to which they are sensitive, or the striped pattern is not of a high enough contrast, then the insects do not detect the movement of the barber's pole stripes and do not control their flight speed and direction accordingly. Hence, it is possible to

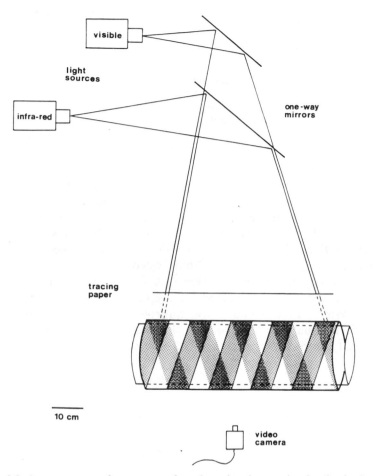

Figure 2.8 Arrangement of apparatus for observing insects in the 'barber's pole' tunnel. Insects are placed in the inner, stationary cylinder. The outer cylinder rotates in either direction, producing the visual impression to a flying insect of a wind in the opposite direction to the apparent stripe movement. The video camera beneath the barber's pole tunnel records the silhouette images of insects against a circle of light from above. Two types of light can be superimposed by the arrangement of light sources and the one-way mirrors as shown. At low visible light intensities, infra-red light was necessary to illuminate the camera (see text).

get a good estimate of difference thresholds for the visual responsiveness of insects in free flight, although the absolute threshold cannot be measured as totally black stripes cannot be produced in this setup. Experimental insects are viewed from below the tunnel as silhouettes against the circle of light coming from above. The tunnel is surrounded by white material from the diffusing screen down, with a hole for the camera. This reduces the possibility of the insects' detecting any fixed visual reference points from other pieces of the apparatus or objects in the room. The main visual input is thus the rotating stripe.

2.4.3 An automated vertical wind tunnel

Flying insects do not always rely on visual feedback from below for sustained flight. The development of vertical wind tunnels that would hold individual winged aphids for extended periods of free flight began in the 1950s (Kennedy and Booth, 1956). Here, the phototactic response of an aphid attracts it towards a bright overhead light while a downdraught of air, balancing out the rate of climb, prevents upward movement. In order to observe the behaviour of individual insects in such chambers, constant attention and the manual adjustment of air flow was required. As recently as 1985 it was thought that such a flight chamber was 'virtually impossible to automate' (Dingle, 1985). However, automation has now been achieved in an updated wind tunnel using real-time video tracking linked via a computer to servocontrol wind flow (Fig. 2.9; David and Hardie, 1988). The positional coordinates of the insect are logged automatically on to disk for later analysis, and the computer also controls the intermittent presentation of plantlike visual cues by the back-illumination, with green light, of a translucent target screen mounted in a side wall. The basics of the flight

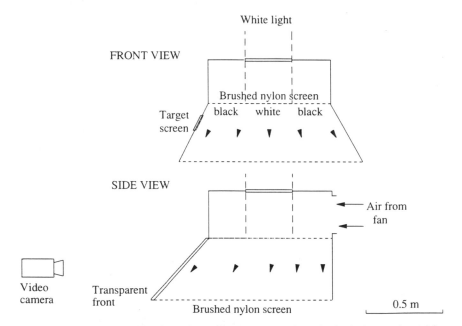

Figure 2.9 Front and side view plans of an automated vertical wind tunnel used for aphid flight studies.

chamber remain similar to earlier designs, but the shape creates an air–velocity gradient with minimal turbulence and dead space. Such a gradient buffers any small changes in the rate of climb of the aphid. The vertical rear wall and sloped front produce a slight horizontal back-to-front component to the wind which ensures that the flying insect faces the rear of the chamber as it tries to reach the overhead light. The front-up/tail-down body axis during flight maximizes the light reflected from the body and wings directly into the video camera, which is mounted as far as possible away from the chamber, to reduce perspective (see Section 2.2.2). As the wingbeat frequency is about 125 Hz, the wings appear continuously extended on the video image, which again enhances visibility. This positioning is highly advantageous, as aphids are small and darkly coloured while the video tracking system requires high contrast between the background and object to be tracked. For this reason, as well as to minimize the visual cues perceived by the flying aphid, the inside of the flight chamber is matt black, with the rear wall covered in black velvet, as this material reflects little light and forms the background to the flying aphid on camera. The chamber floor is black brushed nylon, with the ceiling a brushed nylon screen comprising a central white square with a black border. In addition, contrast between the flying aphid and the rear wall is enhanced by a shadow board which casts shadows over the rear wall.

The video signal is fed through an HVS video tracking unit (HUS Image, Ormond Crescent, Hampton TW12 2TH, UK) which is set to track the highly contrasting white aphid image on the dark background. In the flight chamber, light intensity is greatest – and therefore tracking most efficient – in the central region of the flight chamber beneath the light window, but sufficient light is transmitted through the black material border to illuminate the aphid. The teletracker has a variable window of sensitivity which is adjusted so that, although the complete flight chamber is in camera view, tracking is limited to an area which eliminates bright areas such as the lateral target screen (particularly when green-illuminated). On the floor of the chamber on the right-hand side a small white bead is placed. The coordinates of the bead are known, and if the insect moves outside the tracking window or the tracker loses the aphid, the bead is tracked. Durations of non-insect tracking can then be assessed.

The two-dimensional X-Y coordinates of the flying aphid are fed from the tracker unit to a microcomputer. The vertical position is utilized to servocontrol the air speed via a stepper motor-operated valve on the fan inlet. As the insect moves upward or downward from a narrow horizontal corridor at the level of the target screen, the air speed is altered accordingly so that the insect is returned to target level. When the target is green-illuminated the insect is thus at the same level and, as it is facing the rear of the chamber, this intermittent visual cue is always perceived in the same lateral field of view. Once the coordinates have been logged, analysis software examines for persistent flight towards the overhead light or a response by horizontal movement towards the green-illuminated target screen.

Manually operated flight chambers of this type have been used to study flight in a number of insect species as well as aphids (Blackmer and Phelan, 1991). At present the studies are restricted to weak, day-flying insects with positive phototactic responses, but the opportunity for video-linked automation of wind tunnels now exists.

2.5 SUMMARY

The use of video technology to study the behaviour of insects in free flight offers certain advantages over tethered flight techniques. It may not be possible for all insects, but initial problems may be overcome by attention to the precise sensory stimuli necessary for flight and the relevant illumination for imaging. There can be no doubt that video is a powerful analytical technique in the behavioural sciences.

REFERENCES

Blackmer, J.L. and Phelan, P.L. (1991). Behaviour of *Carpophilus hemipterus* in a vertical flight chamber: transition from phototactic to vegetative orientation. *Entomologia Experimentalis et Applicata*, **58**, 137–148.

Colvin, J., Brady, J. and Gibson, G. (1989). Visually guided, upwind turning behaviour of free-flying tsetse flies in odour-laden wind: a wind-tunnel study. *Physiological Entomology*, **14**, 31–39.

David, C.T. (1982). Compensation for height in the control of groundspeed by *Drosophila* in a new 'barber's pole' wind tunnel. *Journal of Comparative Physiology*, **147**, 485–493.

David, C.T. and Hardie, J. (1988). Visual responses of free-flying, summer and autumn forms of the black bean aphid, *Aphis fabae*. *Physiological Entomology*, **13**, 277–284.

Dingle, H. (1985). Migration, in *Comprehensive Insect Physiology, Biochemistry and Pharmacology*, Vol. 9, eds. G. Kerkut and L.I. Gilbert, Pergamon Press, Oxford, pp. 375–415.

Gibson, G. (1985). Swarming behaviour of the mosquito *Culex pipiens quinquefasciatus*: a quantitative study. *Physiological Entomology*, **10**, 283–296.

Grace, B. and Shipp, J.L. (1988). A laboratory technique for examining the flight activity of insects under controlled environment conditions. *International Journal of Biometeorology*, **32**, 65–69.

Kennedy, J.S. and Booth, C.O. (1956). Reflex and instinct. *Discovery*, **17**, 311–312.

Kennedy, J.S. and Marsh, D. (1974). Pheromone-regulated anemotaxis in flying moths. *Science*, **184**, 999–1001.

Kennedy, J.S., Ludlow, A.R. and Sanders C.J. (1981). Guidance of flying male moths by wind-borne sex pheromone. *Physiological Entomology*, **6**, 395–412.

Miall, R.C. (1978). The flicker fusion frequencies of six laboratory insects, and the response of the compound eye to mains fluorescent 'ripple'. *Physiological Entomology*, **3**, 99–106.

3

Parasites and Predators

M.J. Varley, M.J.W. Copland, S.D. Wratten and M.H. Bowie

3.1 APPLICATIONS OF VIDEO TECHNOLOGY TO PARASITOID AND PREDATOR ECOLOGY

Scanning the literature concerning the ecology and behaviour of parasitoids and predators seems to show that, as yet, few workers are using video techniques in their studies. Indeed, many of the references at the end of this chapter do not specifically mention the use of video in their methodologies. However, there is no doubt that video can be used effectively in such studies, and we intend to give examples of applications where its use would be appropriate.

3.1.1 Locomotion

Studies of locomotion are particularly suited to video work, especially where the locomotion is basically two-dimensional and can be contained within a moderate arena. Bigler *et al.* (1988) used video recordings to trace the walking tracks of female *Trichogramma maidis* wasps to assess the capacity of different strains for host location, and hence efficiency for use in inundative biological control programmes.

The reaction of natural enemies to chemical host cues, and the modification of locomotory behaviour they can cause, have also been studied. Analysis of video recordings and direct observations of the insect parasite *Cotesia rubecula* showed that the wasps' ability to find the obligatory host *Pieris rapae* depended entirely on responses to kairomones produced by the hosts' feeding activities (Nealis, 1986). There have been studies of the effect of kairomones that act over short distances (Bouchard and Cloutier, 1984;

Video Techniques in Animal Ecology and Behaviour. Edited by Stephen D. Wratten. Published in 1993 by Chapman & Hall, London. ISBN 0 412 46640 6

Carter and Dixon, 1984; Vinson *et al.*, 1978) or as contact kairomones
(Lewis and Jones, 1971; Waage, 1978). Short distance-acting kairomones
are sometimes called 'searching stimulants' because they can arrest parasitoids or predators in a patch contaminated by host or prey, or stimulate
increased searching movements which in turn increases the chance of
host/prey location. These searching movements can be defined in relation
to speed of walking (orthokinesis) and amount of turning of the searching insect (klinokinesis) (Kennedy, 1978). Vinson *et al.* (1978) noted an
increase in the frequency of turning by the female parasitoid *Microterys
flavus* when it came into contact with honeydew secreted by the brown
soft scale *Coccus hesperidum* L. The physical structure of the substrate
may also affect the searching efficiency of parasitoids and predators.
Hulspas Jordan and van Lenteren (1978) showed a relationship between
leaf surface and walking speed of *Encarsia formosa*. The rates of parasitism shown by *Trichogramma pretiosum* were shown to vary according
to the density of trichomes on the leaf surface of the plants infested by
the host (Treacy *et al.*, 1984).

3.1.2 Behaviour

Video techniques are not restricted to studies of locomotion. Other aspects
of behaviour can also be studied, from recording the number of visits of an
insect to a certain area, to fine observations of body posture in response to
the presence of prey or the inspection of a possible host. Halsall and
Wratten (1988a) recorded the visits of the carabid beetle *Agonum dorsale* to
sites of high and low aphid infestation, and showed that non-climbing
predators may be important mortality agents for cereal aphids.

3.1.3 Prey detection

Prey detection and recognition have also been studied using video techniques, in carabid beetles (Wheater, 1989) and in the jumping spider *Maevia
inclemens* (Clark and Uetz, 1990).

3.1.4 Flight

Studies of flight also lend themselves to the use of video, although at present
there seems to be little or no work in this area related to insect parasitoids
and predators. Work on other insect groups often involves using elaborate
flight chambers and high-sensitivity video equipment (Grace and Shipp,
1988). Special infra-red sensitive video equipment has also been used in the
field for measuring the flight trajectories of larger insects (Schaefer and Bent,
1984; Riley *et al.*, 1990), although the small size of parasitoids and some
predators may preclude this approach (see Chapter 1).

3.2 LIGHTING

A rule of thumb in video work is 'the more lighting the better', although in insect studies this must be weighed against the conditions needed for the experiment being undertaken.

3.2.1 Heating effects

Tungsten, photoflood and halogen lights produce considerable amounts of heat which, in close proximity to the subject, may cause unacceptably high temperatures within an arena. It is possible to filter out some of the heat energy content of the light with suitably positioned filters.

It is worth noting that insects with dark or black cuticles may absorb more heat than the paler-coloured substrate, and have a body temperature significantly higher than the surrounding ambient. Because of this, special care must be taken when conducting temperature-sensitive experiments, as the results could be influenced by the local warming, even though the measured ambient was satisfactory.

Fluorescent lighting gives off much less heat, but perhaps the best solution is to use a 'cold light' lamp, which has the lighting element contained within a ventilated box, with the light being transmitted down fibre-optic arms to the subject. Very little heat is transferred in this way. This is the lighting system we use for much of our parasitoid work. It has still been necessary to use heat filters between the fibre-optic source and the subject, however, as it was found that even this method of illumination was capable of raising the temperature of a black object significantly above ambient. In one experiment on a small beetle predator, a thermocouple inserted into a dead beetle was used to monitor this local warming effect, and filtration increased until the beetle remained at ambient.

We have constructed heat filters from transparent plastic boxes that have two plastic pipes (one inlet, one outlet) fixed to them. Cold water (several degrees Celcius below ambient temperature) was then passed continuously through the box, which was positioned between the light source and the subject. Energy measurements were taken with a short-wave radiation detector (capable of detecting both visible and infra-red light energy) and the filters were found to reduce the heat content of the lighting by 90%.

3.2.2 Interaction with insect behaviour

Light levels can influence the activity of insects, and where this may be important care must be taken to ensure similar light levels between replicates of an experiment, and also that the light level is suitable for the behaviour under study to take place. Some parasitoids need quite bright conditions before they display typical searching behaviour, and experiments carried out at low light levels would produce misleading results.

3.2.3 Non-visible illumination

Where behaviour is influenced by diurnal variation in light levels, for example in largely nocturnal subjects, lighting must be sufficient to illuminate the subject but not to influence its behaviour. In many cases it is possible to illuminate the subject with a light source which is outside its visual perception range, but still affords good illumination to a suitably sensitive camera. Halsall and Wratten (1988) used two 60 w red tungsten bulbs to illuminate an arena while studying nocturnal carabid beetles. It had earlier been established (Griffiths, 1983) that similar beetles could not perceive red light from this source, so for practical purposes the subjects were considered to be in the dark. Many CCD cameras have a good response to near infra-red light, and so such sources can be used to illuminate nocturnal subjects. True infra-red equipment is considerably more expensive, as are photomultipliers, but these can give true 'night vision' capability.

3.2.4 Directional influence of lighting

Directional influences of lighting must also be considered (Andersen, 1989), and it may be necessary to use some method of diffusion between light source and subject to avoid phototactic movement towards or away from the light. Any diffuser will cut down the amount of light reaching the subject. Diffuser material can be quite simple, e.g. white paper, or translucent white Perspex. If a directional light source is unavoidable, reflectors can be used on the opposite side of the arena to balance the lighting.

3.3 ARENAS

When using video, experimental arena design is necessarily influenced by the limitations imposed by the video equipment. Only two-dimensional arenas can be studied with a simple single camera setup, and the size of the arena is constrained to that which allows the required observations to be made from the resulting recording (see Section 3.4). The arena must also allow adequate lighting to be provided.

3.3.1 Size of arena

There is often a conflict between the desire for a large arena, affording the subject room to move relatively freely, and the need to record behaviour that can only be seen if the subject is observed at close quarters. This is not only a problem with small insects, as larger insects may need very large arenas, which necessitate a wide field of view and subsequent loss of detail of the individual insect. This generally means that if video is to be used for close observation of behaviour, the arena must be relatively small, and for observation of movement, especially of very active insects, the arena must be relatively large. It may be desirable to use the video recording to capture movement information, but to augment this with a spoken commentary describing the fine details of action, such as antennation or preening.

Studies have shown that the effect of arena size varies with the subject species and the behaviour being observed. In an examination of rates of parasitism by several species of the hymenopteran parasitoid *Trichogramma* (Thorpe and Dively, 1985) in various-sized arenas, it was found that one species achieved the highest rates of parasitism in the smallest arena and the poorest rate in the largest, whereas in another species this pattern was reversed. It was concluded that this variation possibly arose from differential vertical stratification of searching behaviour between the two species. Investigation of the predation of the homopteran *Nilaparvata lugens* by the arachnid *Theridion octomaculatum* showed that the functional response model only became a typical sigmoid III type when the space and complexity of the arena was increased (Ge and Chen, 1989). Cave and Gaylor (1989) concluded in their studies of the functional response of the hymenopteran *Telenomus reynoldi* that data obtained from a more complex arena containing an artificial plant might be more appropriate to what actually happened in a field environment. We have also found that functional response data on the parasitoid *Leptomastidea abnormis* observed in a small arena was unsatisfactory.

The arena may be dictated by the experiment itself, for example, the use of an olfactometer to study pheromone or kairomone effects. In this case, the olfactometer is the arena. In other types of study, the design of the arena is more flexible. With studies on host location in parasitoids, clear plastic sandwich boxes or large Petri dishes are often used. Where chemical cues may be present, it must be remembered that small sealed boxes may become saturated with the chemical vapour, and the subject, becoming habituated to the cue, will not respond in the same way as it does in an open situation. This can be circumvented by setting fine gauze or mesh panels into the side of the container to allow gas transfer with the outside environment to take place.

3.3.2 Environment within the arena

It must be borne in mind that the arena has an environment of its own, and its temperature and humidity may be different from the surrounding laboratory. This may be used to advantage, as it is often easier to create the desired environment on the scale of the arena than it is to modify the entire room. Abou-Setta and Childers (1987) used an arena with a water reservoir to rear and observe mites on citrus leaves. A deep Petri dish acted as the reservoir, with a wick leading to another dish containing a cotton-wool disc attached to the wick. Leaves or leaf discs were supported on the cotton wool, and the unit was enclosed by a lid made from an inverted shallow Petri dish, in which holes had been drilled for ventilation. This system maintained a 70–80% relative humidity, which was ideal for the mites, and a mature citrus leaf could last up to 8 weeks within the system. Observations were made through the transparent lid.

Totally sealed arenas, especially those containing moist material, may suffer from condensation, which could trap the observed insects or obscure the view of the observer. It is possible to set up known humidities inside sealed arenas by using reservoirs of various salt solutions. Each salt, when used as a saturated solution, will equilibrate with the air in the chamber and produce a predictable humidity. A range of salts and the corresponding relative humidities above their saturated solutions at 25°C is given in Table 3.1.

Table 3.1 Relative humidities above saturated salt solutions at 15°, 20°, 25° and 30°C

Salt	Relative humidity (%)			
	15°C	*20°C*	*25°C*	*30°C*
Potassium sulphate	97	97	97	96
Potassium nitrate	94	93	92	91
Potassium chloride	87	86	85	85
Ammonium sulphate	81	81	80	80
Sodium chloride	76	76	75	75
Sodium nitrite	—	66	65	63
Ammonium nitrate	69	65	62	59
Sodium dichromate	56	55	54	52
Magnesium nitrate	56	55	53	52
Potassium carbonate	44	44	43	43
Magnesium chloride	34	33	33	33
Potassium acetate	23	23	22	22
Lithium chloride	13	12	12	12
Potassium hydroxide	10	9	8	7

(From *Instruction Sheet – Griffin Choice Chamber*, Griffin and George, Wembley, Middlesex.)

3.3.3 Preventing the subject escaping

If the arena is not sealed, there may be a problem with the subject escaping. Non-flying subjects may be contained within an area by means of a simple wall. To ensure that they cannot climb the wall and escape, it must be made of a material that affords them no grip: Halsall and Wratten (1988) used polythene coated in an aqueous suspension of polytetrafluoroethylene (PTFE) to line hardboard walls which enclosed the arena for a study using carabid beetles.

3.3.4 Behaviour influenced by arena design

Some insects and mites exhibit behaviour which makes them remain in the vicinity of the edge of the arena, and constantly walk around the edge. This 'edge walking' was noted by Berry and Holtzer (1990) when studying predatory mites. The mites used this behaviour in the wild to follow leaf edges and hence disperse to new sites where prey may be found, where they would resume a random searching pattern, but in the circular arenas used this behaviour significantly affected the prey-searching data.

It may not be necessary for the entire arena to be kept under observation. Often interest lies with a small part of the whole arena, where some specific action will take place. For example, a study may look at a single patch of hosts to determine the number of parasitoid visits, or to look at the behaviour of the parasitoid when approaching or encountering the patch. A similar situation may occur with sedentary prey, such as aphids, in predator studies: the prey may be observed continuously but the more active predator can be given a relatively large arena in which to search.

3.3.5 Subject marking

In cases where it is difficult to distinguish the subject from the background (or from other subjects in the same arena) it may be possible to mark the subject in such a way as to make it stand out. Large insects such as beetles can be marked with spots of white or coloured paint applied to the elytra or thorax. Solvent- or oil-based paints may affect the subject, so a water-based paint such as acrylic is preferable. Another marking method is to use a fluorescent paint, and include a proportion of ultraviolet light in the illumination. Very small and active insects are difficult to mark in this way, however, although fluorescent dusts are a possibility.

If a monochrome camera is being used, it is sometimes possible to use coloured filters to enchance the contrast between subject and background. For example, brown insects on green leaves may have very similar tonal values when viewed as shades of grey, but by using a green filter the insects will be considerably darkened with respect to the leaf. Using filters may require the general illumination to be increased. The same effect can be achieved by using coloured light for the illumination.

The use of a contrasting substrate may be possible in cases where this does not affect the behaviour of the subject. Halsall and Wratten (1988a) used silver sand as a substrate for carabid beetles and compared pitfall trapping scores against a natural soil substrate, but found no difference. This technique is not practical in some cases, for example where the substrate is a leaf or leaf disc.

3.4 DATA EXTRACTION FROM VIDEO

Extracting information from the video recording of an experiment can be very simple, for example where event timing is the only information of interest. In most cases, more complex data of both a spatial and a temporal nature are required, and the extraction of this information can be difficult and time-consuming, requiring several examinations of the video recording. We have applied video technology to our own studies of parasitoid and predator behaviour, and it may be helpful to use this as a case study, as many of the techniques are applicable to a wide range of investigations.

One of our main areas of study has been the efficiency with which parasitoids search for, find and deal with the host species. This involves measurements of movement, such as distances, speeds and path tortuosity, as well as examining time partitioning for handling actions like oviposition, host feeding and cleaning after host handling. The studies have taken place over ranges of temperature and at different light levels, and on different substrates used as arenas, so that substrate topography could be included as a factor.

Because the insects studied are relatively small, being 1–2 mm in length, Petri dish arenas or transparent plastic sandwich boxes have been used. The video camera is mounted on a copy stand pointing downwards, with the arena placed on the stand base. The camera height is adjusted to allow the arena to fill as much of the field of view as possible. Lighting is provided by a fibre-optic 'cold light' source. To obtain bright lighting of around 120 W/m^2 up to six fibre-optic sources must be used, and to avoid heat radiation these must have heat filters attached. Most of our video work has been carried out using domestic-level video products.

3.4.1 Acetate sheet

At first we used the somewhat crude technique of taping transparent acetate sheets to the video monitor and, during playback of the recording, tracing the path of the insect by drawing on the acetate with a spirit marker pen. We used a character generator/stopwatch facility attached to the camera to generate a stopwatch on the video picture during recording. Subjects were followed in this way for the length of time decided upon as appropriate by the experimenter, usually 5 or 10 minutes. At the end of the observation, the acetate held a trace of the path of the insect over the observed time period. Other features such as substrate shape (e.g. leaf outlines) and host or host cue positions could then be added to allow simple zoning of the path data.

The information that could be extracted from the traces was mainly distance and tortuosity of path. Distance was measured by following the track with a planimeter of the type used for map work. Tortuosity was more difficult to extract, and the process necessitated breaking the path into many small sections and examining the angles formed between adjoining sections. Speed could only be derived as an average for the total observation, with variations during the observation only possible to extract by repeated playing and timing of sections of the recording. These sorts of data would be taken for parasitoids introduced into arenas containing hosts, containing extracts of hosts (haemolymph or honeydew), and without any host presence or contamination. The observations would be taken at different constant temperatures, under different intensities of illumination, and on different substrates (leaves of different plant species, with either topographical or chemical features that might influence parasitoid searching efficiency).

Time partitioning of behaviour had to be extracted from the recording during a separate playback, using a bank of stopwatches and tally counters.

Obviously, where several parameters (light, temperature, plant species) were being studied, the number of acetates produced was quite large and the time taken to extract the data from them considerable. Because this area of research formed a major part of the group's interest, we decided to develop a technique that would speed up the process of data extraction from video recordings, and at the same time allow us to make much more detailed analysis of the subjects' behaviour. To this end, we turned to the personal computer.

3.4.2 Using computers for data extraction

Such is the progress in electronics and computing that it is now possible to buy extremely sophisticated and powerful personal computer equipment that can take a video picture frame by frame, analyse it for movement, shapes, colours, edges and much more, and record the extracted data, all in real time as the video picture is being received. Such systems are not cheap, and their sophistication requires a considerable amount of learning on the part of the user, so that he or she can configure the system to produce the required data.

What we wanted was a much simpler system: something that could be used by many people from undergraduate to postgraduate level, that would provide the same sort of information as the acetate sheets (or more) but in a much shorter time. We had begun to use computers with graphical user interfaces – instead of typing commands you generally moved a pointer on the screen to point at a graphical representation of a program you wished to run, and pressed a button on the device (called a mouse) that you moved in order to move the pointer. This 'point and click' method of information input offers us the opportunity to use the video screen itself rather than an acetate sheet stuck over it as the medium to take our measurements from. By mixing the image of the pointer (controlled by the mouse) with the image coming from our video source, we can manually move the pointer over the video picture. As the computer knows exactly where the pointer is at any given time, it can perform calculations on the position and movement of the pointer. If the pointer was made to follow a moving object on the video picture, those measurements would reflect the movement of the video object. All these calculations can be done as the video signal is received, so that the data are generated as the movements happen.

Genlocks

The technology needed to get the video picture on to the computer screen is very simple. Rather than the sophisticated digitization technique used by higher-end computing systems, where the video signal is converted into many picture elements or pixels that can then be manipulated and displayed, we use a simple video mixing technique. The incoming video signal and the picture generated by the computer are synchronized by a device called a genlock. Switching between these two pictures results in flawless splicing of the pictures.

In our system the switching is based on the colour occurring on any part the computer display – if the computer is trying to display the target colour, the external video signal is switched in and displayed, with any other colour causing the computer graphics to be switched in and displayed. For example, if the target colour was blue, any occurrence of blue on the computer display would be replaced by the incoming video. A blue circle would appear as a 'hole' in the computer display, with the incoming video visible 'through' the hole. By making sure that most of the computer display consists of the target colour, the incoming video can be made to fill most of the display. The only computer graphic left on the display is the pointer, which appears to 'float' over the video picture.

Genlocks are widely used in professional video applications, and can be relatively expensive if capable of producing output signals of a quality suitable for broadcast. However, most modern personal computers have a range of purpose-built genlocks which are aimed at the amateur market and are reasonably priced. As our application is not concerned with rerecording the mixed video, all we are concerned about is the quality of the picture on the monitor. Genlocks capable of providing RGB signals for the computer display give much better definition than those outputting purely composite signals. The comments made earlier about compatibility within video systems apply to genlocks as well, as the different video standards all have slightly different synchronization patterns.

In most cases the genlock will use the synchronization signals of the incoming video as a basis for locking the video and computer display together. In the Commodore Amiga computer, because the computer display is already generated with standard video synchronization, all the genlock has to do to lock the signals is to take control of the computer's internal timing and move it until the signals match. In other computer systems where the display is not generated with standard video synchronization, much more work must be done by the genlock, converting the computer display to standard video before it can lock to the incoming signal. This makes the Amiga genlocks much simpler and therefore cheaper.

Timebase correction

It follows that if the source video is being used for synchronizing the two signals, any instability in the source signal will cause problems with the mixing. The sync signals produced by VCRs are noticeably worse than those produced by cameras, and it is sometimes the case that using a VCR as the source video produces unstable mixing in the genlock. This manifests itself as 'roll' or 'jitter' of the mixed picture, which for our purposes would make the system difficult or impossible to use. If the sync problem from the VCR is purely one of signal quality, the sync signals can be cleaned up or regenerated by an electronic device which takes the original video signal, breaks it down into the component sync and picture signals, and replaces the sync signals with good ones before reassembling the complete video signal. Such pieces of equipment are very cheap. However, if the poor synchronization signal is due to a manipulation of the tape speed, as is the case with slow-motion, time-lapse or still-frame features, the sync signals are too badly disrupted to allow stable genlocking. In these cases, another method must be used to resynchronize the video signal. Unfortunately, this is not such a simple process, and can only be achieved by a timebase corrector that works as a field or frame store. Put simply, every frame of the video is captured in a memory and then re-output at a very stable frame rate with reconstructed sync information. Such equipment is relatively expensive, being used primarily in the professional video industry, although cheaper units are becoming available due to the growing hobby video market. Timebase correctors are even being fitted as standard to some new domestic VCRs.

We have used Panasonic editing desks, which have integral timebase correctors on both input channels. Despite having many other functions and facilities for both audio and video mixing, these units are still cheaper than dedicated timebase correctors.

Typical equipment

A typical basic system would consist of a video camera, a video recorder, a computer (in our case a Commodore Amiga) and a genlock. A copy stand or tripod and a light source would also be required for most applications. If a time-lapse or slow-motion (i.e. variable recording speed or playback) recorder is being used, a timebase corrector must be placed between recorder and genlock. A printer would allow hard copy of the data to be produced. A typical system layout is shown in Fig. 3.1.

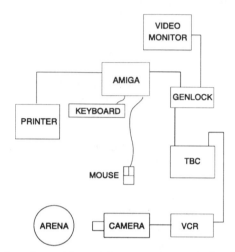

Figure 3.1 Typical equipment layout for computer analysis of video recordings.

Software

Once the computer graphics and video source are mixed satisfactorily, all that remains is to produce a computer program to allow movements of the pointer to be translated into measurements relative to the video source. As no such software existed, we have written our own. We have developed software on the Amiga platform because of the ease of using that machine with video equipment, and also because its multitasking operating system (capable of allowing many tasks to be carried out by the computer simultaneously) and graphical user interface (with mouse-driven pointer) allowed the software to be sophisticated yet user-friendly.

3.5 MICROMEASURE – VIDEO ANALYSIS SOFTWARE

The initial aim of our software development was to provide a replacement for the acetate sheet method of data extraction. In practice, the implementation of a measuring system based on the movement of a mouse-driven pointer will vary depending on the microcomputer platform chosen. On the Amiga platform, movements of the mouse (and hence the pointer), along with its position and the time of each movement (accurate to milliseconds) is available to the programmer via the internal workings of the graphical user interface, which is called Intuition.

In a multitasking system like the Amiga's, the operating system keeps track of the individual applications running at any time. If an application requests it, the operating system will inform the application of various events perceived by the operating system. Such events can include the movement of the mouse, mouse buttons being pressed and released, keys on the keyboard being pressed and released, and so on. The application is informed by a message being sent to it, and each such message contains timing and positional information. It is fairly easy to comprehend that if an application receives messages about the pointer position each time the pointer moves, it is a simple matter to totalize the distances of each move. The messages contain exact timing information, and are 'queued'. This means that the application need not itself be performing calculations in real time, as it can work through the queue of events at its own pace. The fact that the pointer movements (and other input events) are dealt with by a part of the operating system running simultaneously with the measuring application is made possible by the multitasking capability of the machine. This incidentally makes the pointer movements very smooth, which helps the user to achieve accuracy when measuring.

3.5.1 Calibration of measurements

Distance calibration

The pointer positions sent to the measuring program are in the form of screen coordinates. To obtain real measurements (in relation to the video picture being analysed) from these coordinates requires a calibration technique. The screen coordinates are sent as x and y pixel positions, in this case with the top left of the picture being 0,0. The screen is divided up into 640 vertical columns and 256 horizontal rows, and at each intersection of a row and a column is a pixel. The resolution ties in well with the approximate resolution of the PAL video picture referred to earlier.

Moving the pointer across the screen causes the active point of the pointer (sometimes called a 'hot spot') to traverse pixels. The operating system scans the mouse for movement on a regular basis, but the exact timing varies depending on the task switching load (that is, the number of programs operating under the multitasking regime). For our purposes, as the pointer is moved relatively slowly, we can assume that a message is sent out almost every time it crosses a pixel boundary into another pixel.

The technique used for transposing pixel coordinates to real units relevant to the video picture relies on having an object of known dimensions in the picture. This can be a marked rule, a sheet of squared paper – literally anything of known dimensions. The calibration routine requires the pointer to be positioned over one end of the known dimension, and a mouse button to be clicked to register the position. The pointer is then positioned over the other end of the dimension and the mouse button clicked again to register the second point. The dimension (in real units) is then typed in. The program can now relate its pixel positions to real units, and data are presented in real units. The calibration object can lie in any orientation, as a correction is made for aspect ratio. This is necessary if the pixels are not square, in which case a vertical pixel offset would not move the pointer the same distance as a horizontal pixel offset.

Once the calibration has been made, the program returns all data in the calibrated units. The calibration factor can be saved on disk (along with other setup information) and recalled should the same configuration be used again.

Time calibration

Time can also be calibrated to allow for slow-motion or time-lapse video recordings to be analysed correctly for time-related data. The process is similar to distance calibration, but in this case the start and stop clicks are made at the beginning and end of a known time period on the video recording. This is usually provided by a stopwatch display on the video itself.

3.5.2 Taking measurements

Once the system is calibrated to the video picture, measurements can be made. A subject is introduced to the arena and allowed to acclimatize if necessary. When the observation is about to begin, the pointer is guided over the subject by moving the mouse. A click of one of the buttons on the mouse begins the observation. The observer must now maintain the position of the pointer above the subject throughout the observation period.

During the observation the keyboard becomes an event recorder. Any key pressed will be noted by the program, and the time between pressing and release will be added to an accumulator for that key. Using this feature, simple time partitioning can be achieved.

When the observation period is over, a click on the mouse button signals the program to stop computing the information from the mouse movements. The summary of collected data is available for examination on the screen, and can be printed out on to paper and written to a file on magnetic disk for later processing. The measurements available are:

- Distance – the total distance travelled by the subject over the course of the observation;
- Duration of observation;
- Time during observation that the subject was moving or stationary;
- Gross speed – distance/time (in seconds) to give average speed throughout the observation;
- Net speed – average speed subject travelled at while in motion;
- Tortuosity of path relative to unit distance;
- Tortuosity of path relative to unit time;
- Event table containing the identity of keys pressed during observation, the number of times each key was pressed, and the total time that each key was held down for.

3.5.3 Partitioning data between zones

In many studies there may be definite zones within the arena where behaviour may be expected to differ, and comparison of data obtained from the different zones is desirable. The program accommodates this need by allowing the user to define areas within the field of view as distinct zones, the data from which will be held separately. This is achieved by setting up an area of memory (or bitmap) which acts as a duplicate of the visible computer display, having the same pixel resolution. By allowing the user to 'draw' zones on to this duplicate display and fill each zone with a separate colour, all the program has to do during an observation is to check the colour of the 'zone display' relative to the pointer position on the measuring display, and store the data accordingly. The actual zones are not visible during observation, to minimize any bias the user might show if the subject is inside or outside a designated zone.

A typical use for this zoning feature would include studies of parasitoid searching behaviour in relation to host or host cue presence. The host or patch of hosts could be contained within a zone denoting the contamination of the arena by their honeydew or waste, and the behaviour of the parasitoid within this zone (and therefore its awareness of the presence of hosts in the vicinity) compared with the behaviour in areas not indicating any host presence.

Extremely complex zoning can be achieved using this method, for example a series of concentric rings of known diameters radiating from a point. Using angle measurement of the track referenced to the central point (which gives a measure of how directly the track approaches the point), a measure could be made of the distance at which a parasitoid or predator recognizes the host or prey. The data partitioned in this way can be viewed zone by zone, or as a summary of the data across all zones (i.e. as if no zones existed).

3.6 MICROMEASURE IN USE

Hag Ahmed (1990) used MicroMeasure to study the searching behaviour of the parasitoid *Aphidius matricariae* Haliday in relation to plant structure, leaf topology and host cues. He recorded the time spent searching, the speed of walking and the rate of turning for the parasitoid when presented with leaves showing differing surface topology. Table 3.2 shows the results of this study.

A JVC GX-N7E colour video camera and a JVC HR-S10EK video recorder were used to record the activity of parasitoids in an 88 mm Petri dish which contained the selected leaves. The camera was positioned on a copy stand directly above the Petri dish. A Schott KL1500-T cold light source was used to illuminate the arena, producing light levels of about 45 W/m^2. Single female parasitoids were introduced to the arena and their movements recorded for about 5 minutes. All experiments were conducted at room temperature, which was $20°C \pm 4°C$.

Table 3.2 Searching behaviour component of *Aphidius matricariae* on lower surfaces of different leaves ($n = 15$)

Plant	Time searching (s) (mean) NS	Speed (mm/s) (mean)	Turns >75° (mean in 5 min)
Potato	23.19	2.75 c	6.06 a
Turnip	21.25	3.75 bc	5.07 ab
Peach	22.52	3.54 c	4.8 ab
Cauliflower	20.05	3.65 bc	5.13 ab
Lemon	7.58	4.75 ab	2.2 c
Faba bean	15.3	5.31 a	3.4 bc

Means within columns sharing same letters are not significantly different at the 5% level. NS = no significant difference at the 5% level.

The recordings were then analysed with MicroMeasure by connecting the video recorder to the computer via a 'SuperPic' genlock. In the version of MicroMeasure used for this study, a user-selected angle is entered, change in direction greater than which would result in the program counting a 'turn'. In this study, changes in direction greater than 75° were counted as turns. Later versions of the software use a slightly different measurement for tortuosity, allowing accumulated degrees to be recorded over unit distance and unit time, with independent user-settable sampling intervals for both measurements.

Hag Ahmed also measured the tortuosity of the parasitoids' path (number of turns over a set observation time) when presented with clean leaves, leaves contaminated with host honeydew, and leaves where hosts were actually present. Typical traces obtained are shown in Fig. 3.2.

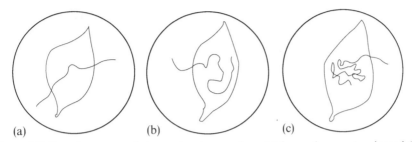

(a)   (b)   (c)

Figure 3.2 Search pattern of *A. matricariae* on clean (a) honeydew-contaminated (b) and aphid-infested (c) leaves.

3.7 OTHER VIDEO ANALYSES OF PREDATION

There have been other recent attempts to quantify predation, or at least the movements of predators, and these have involved systems other than the glasshouse. The examples to be considered now are:

- New Zealand work which tracks automatically the movements of predatory and prey spider mites;
- work in the UK which has used video in the laboratory to evaluate the efficiency of insect 'pitfall traps';
- video systems in cereal fields to evaluate the searching behaviour of predatory beetles on the soil surface.

3.7.1 Automatic tracking of small invertebrates in the laboratory
(Bowie & Worner, 1992; Bowie, 1992)

This technique was developed for spider mites, which are one of the most difficult arthropod groups to study because of their small size (0.2×0.4 mm), difficulty of handling and high mobility. The versatile system described here can track automatically and record a two-dimensional track at predetermined time intervals as frequent as 1 s. The program produces a series of coordinates over the sampling period at a defined interval. When the coordinates are sequentially joined a track is formed which closely models the path of a mite. Analysis of sequential coordinates generates information on activity, directionality, speed, distance and tortuosity. Examples of computer output and motion analysis are given for two mite species, *Typhlodromus pyri*, a predatory mite, and the two-spotted spider mite (TSM) *Tetranychus urticae*, a prey species of *T. pyri*. The system can also be used for other invertebrates.

Arena 1: Agar/leaf disc method for two-spotted mite (TSM)

Healthy mated female mites were placed on to a 24 mm leaf disc which was placed abaxial side up on semisoft agar (0.1% w/v) in a 50 × 9 mm Petri dish base. The reasons for using semisoft agar were to create a barrier to contain the mites; to keep the leaf disc fresh; and to allow light from below to silhouette the mite, creating an image with good contrast.

Arena 2: Tack-trap/coverslip method for T. pyri

A 16 mm diameter microscope slide coverslip was washed in distilled water and, when dry, placed on a thin layer of Tack-trap (a sticky barrier) in a 50 × 9 mm Petri dish base. The base was glued inside a 90 mm diameter Petri dish and a tight-fitting lid was added. By using saturated salt solutions in the outer Petri dish as a surrounding 'moat', a humid micro-environment was created.

The light source used was a Schott KL1500 fibre-optic cold light with a terminal ring illuminator that produced even lighting through an opalized sheet of acrylic. An earthed aluminium spacer was used between the Petri dish and the acrylic to avoid static charges which influence mite movement. The apparatus was enclosed in a 400 × 400 × 400 mm acrylic blackout box to prevent wind currents and directional light influencing the mite behaviour. A monochrome Burle (TC304EX) low-light high-resolution ($\frac{2}{3}$ in format CCD) video camera fitted with a 55 mm Micro-Nikkor lens was mounted over the blackout box with the lens protruding into it. The images were recorded using a Hitachi DA4 video cassette recorder (VCR) and viewed using an AWA Colortrack monitor.

Mite behaviour was analysed in real time or recorded on to a TEAC E-180HX cassette tape which was then played back into the image analysis system. Images were input into the system through an IVT-9SP digital time-base corrector (TBC) which electronically corrected the improper tuning relationships of the sync signals created by mechanical and electronic errors in the VCR. The signal was then fed via an RGB transcoder into the Magiscan image analysis coprocessor. The system was 'driven' through menus by a 386 computer and a mouse (Logitech). The image screen was a 512 × 512 pixel format. Figure 3.3 illustrates the linkages between hardware components.

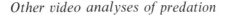

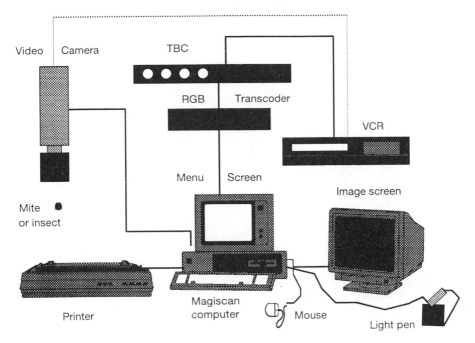

Figure 3.3 Diagrammatic representation of image analysis setup.

Image processing

A Magiscan computer (Joyce Loebl) digitized the images using the BUGSY program. A threshold was set using a mouse to separate the mite image from the background. The image quality was optimized by adjusting the video level and black level on the TBC. Calibration of a known distance was entered using the Joyce Loebl light pen and the scale factor calculated was then used in the analysis. The frequency of the sampling interval was also entered (1 per second).

Once analysis was initiated, the mite's x and y coordinates were automatically recorded and joined in real time to produce a track which models the movement of the mite. The sequential x, y data generated were viewed at the termination of each run. Binary images of tracks were edited when required to obtain standard sample lengths or to isolate behaviours of interest. Both binary images and data were saved on discs for future reference or for additional analysis with statistical packages. A separate FORTRAN program was written to analyse x, y coordinates and to compute required parameters. Single pixel jitter, a problem found by other authors (Mueller-Beilschmidt and Hoy, 1987), was compensated for in the software.

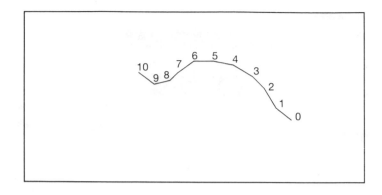

THIS FILE IS: 200792DAT
SCALE FACTOR: 3,000E-02
TIME IS EXPRESSED IN: SECONDS

Index	X	Y	Dist.	Angle	Time	Turn	Speed	Tdist.	Digtime
0	435	311			13:52:41:54				
1	408	344	1.28	120.29	13:52:42:48	−129.29	1.361	1.28	0.94
2	396	388	1.37	105.26	13:52:43:46	24.03	1.396	2.65	0.98
3	381	420	1.06	115.11	13:52:44:40	−9.86	1.128	3.71	0.94
4	344	445	1.34	145.95	13:52:45:39	−30.84	1.353	5.05	0.99
5	304	454	1.23	167.32	13:52:46:32	−21.37	1.323	6.28	0.93
6	261	452	1.29	182.66	13:52:47:31	−15.34	1.304	7.57	0.99
7	230	430	1.14	215.36	13:52:48:24	−32.70	1.226	8.71	0.93
8	208	412	0.85	219.29	13:52:49:23	−3.93	0.861	9.56	0.99
9	168	405	1.22	189.93	13:52:50:17	29.36	1.296	10.78	0.94
10	137	435	1.29	135.94	13:52:51:15	53.99	1.321	12.07	0.98

TABLE FILE IS: 200792.DAT

Mean speed per second	=	1.26
Stationary time (s)	=	0.00
Mean X (Xbar)	=	−0.08
Mean Y (Ybar)	=	0.08
Mean angle	=	136.88
Weighted mean vector length (r)	=	0.11
Arithmetic mean vector length	=	1.21
Mean angular deviation (s)	=	76.28
Absolute mean turn	=	35.07
Mean meander	=	27.93
Distance *as the crow flies*	=	9.68
Index of straightness	=	0.80
Total distance travelled	=	12.07

Figure 3.4 Sample of digitized mite track with ten data points and associated image analysis output.

(a)

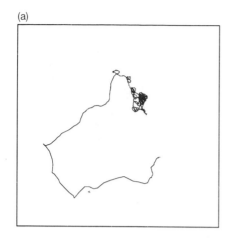

(b)

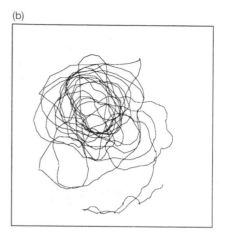

TABLE FILE IS: 021190MTAB

Mean speed per second	=	0.11
Stationary time (s)	=	369.21
Mean X (Xbar)	=	0.04
Mean Y (Ybar)	=	0.18
Mean angle	=	76.51
Weighted mean vector length (r) =		0.18
Arithmetic mean vector length	=	0.10
Mean angular deviation (s)	=	73.15
Absolute mean turn	=	25.49
Mean meander	=	197.32
Distance *as the crow flies*	=	7.86
Index of straightness	=	0.08
Total distance travelled	=	104.23

TABLE FILE IS: 051190ATAB

Mean speed per second	=	0.43
Stationary time (s)	=	253.80
Mean X (Xbar)	=	0.00
Mean Y (Ybar)	=	0.05
Mean angle	=	83.94
Weighted mean vector length (r) =		0.04
Arithmetic mean vector length	=	0.41
Mean angular deviation (s)	=	79.20
Absolute mean turn	=	17.46
Mean meander	=	67.99
Distance *as the crow flies*	=	19.44
Index of straightness	=	0.05
Total distance travelled	=	412.32

Figure 3.5 Examples of TSM behavioural tracks with statistical information. Both mites were subjected to the same conditions (100% RH and 25°C). The variability between the observed behaviours illustrates the need for quantification.

Analysis

Examples of motion analysis are given in Figs 3.4, 3.5 and 3.6 and the movement parameters are described as follows:

1. *Between coordinates*:

(a) *distance travelled* – distance moved per specified time (mm);
(b) *turning rate* – angular deviation from previous direction. A clockwise turn is positive and a counterclockwise, negative (−180 to 180°);

(a)

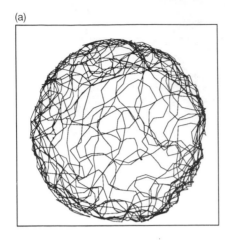

(b)

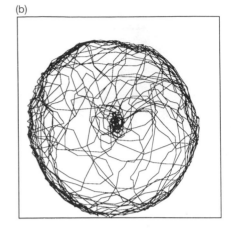

TABLE FILE IS: 081191BTAB   TABLE FILE IS: 120891A1TAB

Mean speed per second	=	0.35	Mean speed per second	=	0.41
Stationary time (s)	=	1365.49	Stationary time (s)	=	1047.27
Mean X (Xbar)	=	−0.03	Mean X (Xbar)	=	−0.02
Mean Y (Ybar)	=	0.22	Mean Y (Ybar)	=	0.05
Mean angle	=	98.01	Mean angle	=	69.98
Weighted mean vector length (r) =		0.23	Weighted mean vector length (r) =		0.05
Arithmetic mean vector length	=	0.34	Arithmetic mean vector length	=	0.40
Mean angular deviation (s)	=	71.24	Mean angular deviation (s)	=	78.86
Absolute mean turn	=	24.05	Absolute mean turn	=	27.85
Mean meander	=	145.98	Mean meander	=	185.55
Distance *as the crow flies*	=	5.15	Distance *as the crow flies*	=	5.47
Index of straightness	=	0.01	Index of straightness	=	0.00
Total distance travelled	=	1019.63	Total distance travelled	=	1209.67

Figure 3.6 Two examples of *T. pyri* searching behaviour with associated statistical information: (a) without prey eggs in arena; (b) with ten TSM eggs in the centre of the arena.

(c) *heading angle* – gives direction on a 360° scale determined by a line joining the last two coordinates;

(d) *Speed* – distance/time (mm/s).

2. *Over entire track*:

(a) *Cumulative distance travelled* – a running total of distance is calculated (mm);

(b) *mean speed* – average rate of movement (mm/s);

(c) *stationary and moving times* – periods of activity and inactivity (measured in seconds);

(d) *mean heading angle* – indicates 'preferred' direction (360° scale);
(e) *mean distance* – arithmetic mean vector length (mm);
(f) *mean angular deviation (s)* – measures clustering or dispersion of heading angle values around the mean (expressed in degrees);
(g) *absolute mean turn* – average of the absolute turn values expressed in degrees;
(h) *mean meander* – average of turn/distance at each sampling point (degrees turned/mm moved);
(i) *index of straightness* – also known as linearity of travel or coefficient of a straight line. This is calculated by dividing the beeline (shortest distance between the start and end points) by the actual length of the pathway, where one is a straight line and values approaching zero are tortuous;
(j) *weighted mean vector length (r)* – describes concentration of sample points around mean direction. A value of 0 indicates no cluster and 1 indicates no deviation.

Several authors have used locomotory activity in mites as a parameter to describe behavioural changes induced by environmental factors. For example, Penman and Chapman (1980) found that three species of mite showed varying responses to different constant temperatures but humidity had little influence on activity. Hirano (1987) found that locomotory patterns of the carmine spider mite *Tetranychus cinnabarinus* treated with fenpropathrin, a synthetic pyrethroid insecticide, differed from control treatments, indicating a behavioural change. Both studies highlighted the influence of environmental factors on mite behaviour but lack detailed description of behavioural parameters. Mueller-Beilschmidt and Hoy (1987) described mite behaviour using computerized tracking, but the only quantitative parameter assessed was rate of activity. Clearly, other descriptors of movement are necessary to assess such a complex factor as locomotory behaviour.

Berry and Holtzer (1990) painstakingly traced the walking paths of predator mites and used the digitized values to calculate turning angles, walking speeds and turning rates. Results showed that two distinct patterns exist: edge-walking and random-walk search. Kitching and Zalucki (1982) suggested that measurements of six components were sufficient to describe the movement path of an animal: (1) the mean and (2) the variance of the angle turned at each step; (3) the mean and (4) the variance of the movement speed; (5) the initial angle of bearing; and (6) the proportion of time spent moving. The method described here provides a means for the assessment of mite behaviour based on these six parameters, and enables the calculation of additional parameters. The speed and efficiency of this technique allows the replication of data, which would be almost impossible by earlier manual methods. It must be stressed, however, that image analysis is not an exclusive substitute for observation when studying behaviour: the observant experimental biologist can still produce insights not possible when video images are scanned. This principle applies to all the case studies in this book.

The video tracking and image processing methodology described above has been used to investigate the effects of a pesticide on mite behaviour; in this case, there was a control treatment replicated with nine adult female mites, and the pesticide (esfenvalerate) treatment with ten. Data output from image analysis were entered into the statistical package Minitab to generate analysis of variance and to obtain frequency distribution information. Turn rate, the mean absolute angles turned and the turn bias (the mean tendency for the mite to turn in a clockwise or counterclockwise direction) were both assessed using analysis of variance. These data were then plotted using Harvard Graphics. Examples of image analysis tracking of mites on control and treated arenas are shown in Fig. 3.7.

(a) (b)

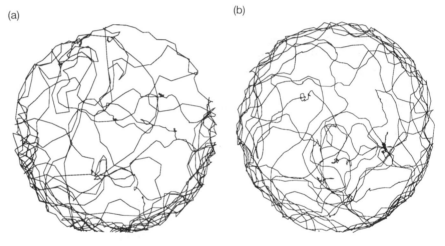

Figure 3.7 Two examples of *T. pyri* locomotory behaviour: (a) a control arena with water on each half; (b) a half-treated arena where the right half is treated with esfenvalerate.

Percentage distributions of distance travelled by *T. pyri* are shown in Fig. 3.8. The two-sample Kolmogorov–Smirnov test was used to compare between distances travelled over the whole arena of control (a) and half-treated arenas (b). This showed that mites on the control arena travelled significantly greater distances than mites on the half-treated arena ($P < 0.0001$).

3.7.2 Video analysis of pitfall trap efficiency

Pitfall traps have been used extensively to monitor the number and activity of surface-active invertebrates, especially Coleoptera and Araneida. Catches

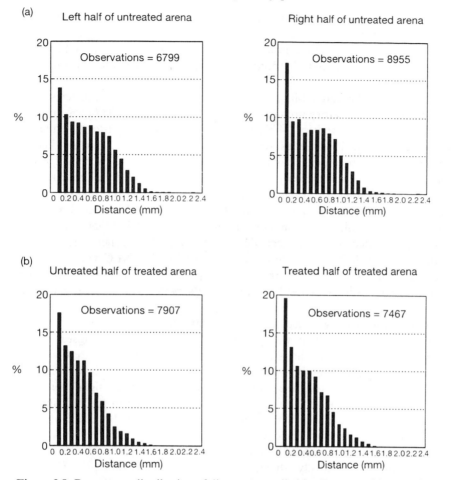

Figure 3.8 Percentage distribution of distance travelled by *T. pyri* on (a) control, and (b) half-treated, arenas.

by these traps depend on the population size, locomotor activity (Mitchell, 1963; Greenslade, 1964) and susceptibility to capture of the animals. Therefore, numbers caught represent only relative numbers of the epigeal fauna present. However, because pitfall traps are cheap and require little labour, they have been used as a major sampling method by, for example, workers studying the epigeal fauna present in agricultural fields (Potts and Vickerman, 1974; Dunning *et al.*, 1975; Sunderland, 1975; Edwards *et al.*, 1978; Sunderland and Vickerman, 1980; Byran and Wratten, 1984).

The use of pitfall traps to compare species' activities relies on the assumption that every species has the same chance of being captured. However, little work has been published to investigate this assumption. Related work includes that of Luff (1975), who demonstrated the differing efficiencies of several types of traps; Greenslade (1964) demonstrated that differently set traps gave differing efficiencies; Obrtel (1971) maximized capture per unit area by manipulating the numbers of traps, and Chiverton (1984) demonstrated that the content of the gut affected the capture of a carabid species.

To compare the 'catchability' of beetle species, seven carabid species were evaluated in a laboratory arena (see Halsall and Wratten, 1988a for details). All beetles were starved for 24 h before the experiment in order to enhance foraging activity. The beetles were introduced into the arena 2 h before the start, and subsequent activity was monitored over a 24 h period using either VHS or U-Matic cassette recorders set in the 24 h time-lapse record mode. The equipment consisted of a National WV 1800B Vidicon monochrome video camera with a Fujinon CF 125C, 12.5 mm F1.4 lens; a Melford D01-17 high-resolution monochrome monitor; an NEC 9507 U-Matic time-lapse video cassette recorder with time–date generator; and a Panasonic NV-8050 VHS video cassette recorder. The NEC video cassette recorder was used only in the early experiments. After this time the Panasonic recorder was used, because of its better image quality during playback.

Some illumination was necessary in the 'dark' period to enable the camera to produce an image. This was provided by two 60 W red tungsten bulbs positioned 60 cm from the surface of the arena. These created a light intensity of approximately $3 \mu E\, m^{-2}/s$ at the substrate surface. Griffiths (1983) demonstrated that *A. dorsale* could not perceive red light from this source, so it was assumed that the species used behaved in the red light as they would in the dark. The camera was positioned 100 cm from the substrate surface, giving a field view of 45×59 cm.

Initial reviewing of each 24 h tape revealed that *Notiophilus biguttatus* was active almost exclusively during the day. *Agonum dorsale, Calathus melanocephalus, Demetrias atricapillus* and *Nebria brevicollis* were almost exclusively nocturnal, and *Calathus fuscipes* and *Trechus quadristiatus* were active throughout the day and night.

The results given in Fig. 3.9 represent the capture efficiency (I), avoidance behaviours (II) and speeds of movement (III) for the four experiments. For the nocturnal species, results presented are from activity in the dark period only. This was because the number of encounters in the light period was very low.

The proportion of encounters with the pitfall traps resulting in capture ('efficiency') was low for all species throughout the experiments (Fig. 3.9). The greatest efficiency was 0.44 recorded for *N. brevicollis* in experiment 3. This experiment was conducted with spring/summer-caught animals on a sand substrate with the lipless trap type. The lowest efficiency achieved was 0, recorded for *D. atricapillus* in experiment 4. This experiment used spring/summer-caught animals with lipped traps set in a sand substrate.

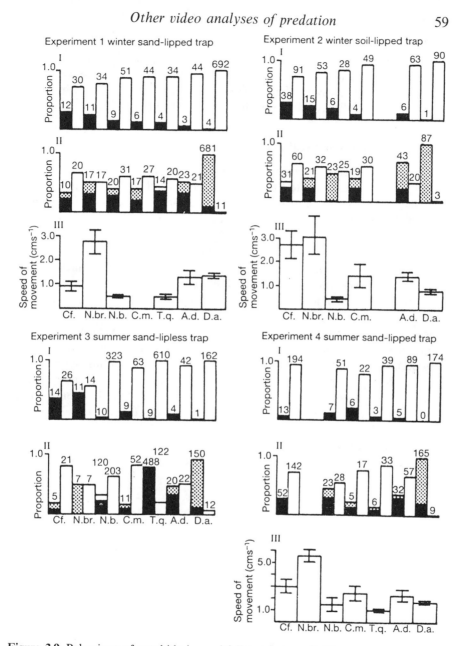

Figure 3.9 Behaviour of carabids in a pitfall-trap arena. (I) The capture efficiency (proportion capture (■) to proportion non-capture (□)); (II) avoidance behaviour (proportion investigate and skirt (■)/skirt (▦) to proportion investigate then retreat (□)); (III) speed of movement (given with 95% confidence limits). Sample sizes are given above each column in observations of types I and II. C.f., *Calathus fuscipes*; N.br., *Nebria brevicollis*; N.b., *Notiophilus biguttatus*; C.m., *Calathus melanocephalus*; T.q., *Trechus quadristriatus*; A.d., *Agonum dorsale*; D.a., *Demetrias atricapillus*.

These results demonstate what was suspected, i.e. that pitfall traps are unreliable measures of activity. However, as long as they are used with caution they do have a 'survey' value in fieldwork, but cannot be assumed to provide density measures. The experiments described above add individual species' 'catchability' to the other limitations of pitfall traps.

3.7.3 Use of video in cereal fields

This was a challenge because of the logistics involved. The site was remote from power and exposed to the elements (see Halsall and Wratten, 1988b for details). Predators often respond to aggregations or 'patches' of their prey, but the behavioural basis of this is not often quantified, especially in realistic situations. In this work, areas of high aphid density were created by artificially infesting a caged area of the crop. The aphids *Sitobion avenae* (F.) and *Metopolophium dirhodum* (Wlk.), obtained from a natural infestation in another field, were introduced into a $2 \times 2 \times 2$ m field cage on alternate days, starting from the middle of May. Two video cameras were positioned approximately 5 m from a hedgerow and 7 m apart, in the crop. Each of these was focused on a patch of moistened silver sand measuring 9×11 cm. The activity of soil-surface predators was monitored over a 3-week recording period using a time-lapse video recorder set in the 48 h time-lapse record mode. The equipment was a JVC 9000 U time-lapse video cassette recorder, a JVC VM-14 psn(G) video monitor, two JVC TKN10 U monochrome Newvicon video cameras, two 8.5 mm fixed-focus auto-iris lenses, two JVC TK U800 U remote control units, a modified JVC SW 108E sequential switcher and an RM-G90 U remote control unit. The switcher was set to give alternate recordings every 5 min from each of the cameras. Illumination

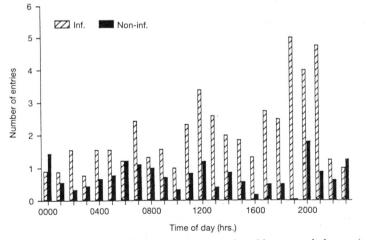

Figure 3.10 The number of beetle entries into the video-recorded prey 'patch' and control areas.

for the night-time recording was provided by two 100 W infra-red light sources, one for each camera. A 1200X Honda generator was used to power all the equipment.

Examples of the output for a group of small carabid beetles (three species) are shown in Fig. 3.10, which shows the higher activity rate of the predators in the 'patch' area compared with that in 'background' prey density areas.

The case studies in this chapter demonstrate the abundance of video hardware and software currently available for work of this type. Labour inputs can still be high, but without such equipment, studies such as the 3-week monitoring of predation in the field (Halsall & Wratten, 1988b) would not have been possible.

(Amiga is a registered trademark of Commodore-Amiga, Inc.)

REFERENCES

Abou-Setta, M.M. and Childers, C.C. (1987). A modified leaf arena technique for rearing phytoseiid or tetranychid mites for biological studies. *Florida Entomologist*, **70**, 245–248.

Andersen, J. (1989). Photoresponse of carabid beetles depends on experimental design. *Oikos*, **54**, 195–200.

Berry, J.S. and Holtzer, T.O. (1990). Ambulatory dispersal behaviour of *Neoseiulus fallacis* (Acarina: Phytoseiidae) in relation to prey density and temperature. *Experimental and Applied Acarology*, **8**, 253–274.

Bigler, F. Bieri, M. Fritschy, A. and Seidel, K. (1988). Variation in locomotion between laboratory strains of *Trichogramma maidis*. *Entomologia Experimentalis et Applicata*, **49**, 283–290.

Bouchard, Y. and Cloutier, C. (1984). Honeydew as a source of host-searching kairomones for the aphid parasitoid *Aphidius nigripes* (Hymenoptera: Aphidiidae). *Canadian Journal of Zoology*, **62**, 1513–1520.

Bowie, M.H. (1992). Using image analysis to study the effect of esfenvalerate on predatory mite behaviour. *Proceedings of the 7th New Zealand Image Processing Workshop 26–28 August 1992*, Christchurch-Lincoln.

Bowie, M.H. and Worner, S.P. (1992). Quantifying mite movement using image analysis. *Proceedings of the 7th New Zealand Image Processing Workshop 26–28 August 1992*, Christchurch-Lincoln.

Bryan, K.M. and Wratten, S.D. (1984). The responses of polyphagous predators to prey spatial heterogeneity; aggregation by carabid and staphylinid beetles to their cereal aphid prey. *Ecological Entomology* **9**, 215–259.

Carter, M.C. and Dixon, A.F.G. (1984). Honeydew: an arrestment stimulus for coccinellids. *Ecological Entomology*, **9**, 383–387.

Cave, R.D. and Gaylor, M.J. (1989). Functional response of *Telenomus reynoldi* (Hym: Scelionidae) at five constant temperatures in an artificial plant arena. *Entomophaga*, **34**, 3–10.

Chiverton, P.A. (1984). Pitfall trap catches of the carabid beetle *Pterostichus melanarius* in relation to gut contents and prey densities, in insecticide treated and untreated spring barley. *Entomologia Experimentalis et Applicata*, **36**, 23–30.

Clark, D.L. and Uetz, G.W. (1990). Video image recognition by the jumping spider *Maevia inclemens* (Araneae: Salticidae). *Animal Behaviour*, **40**, 884–890.

Dunning, R.A., Baker, A.N. and Windley, R.F. (1975). Carabids in sugar beet crops and their possible role as aphid predators. *Annals of Applied Biology*, **80**, 125–128.

Edwards, C.A., Parsons, N., George, K.S. and Heilbroon, T. (1978). Carabids as predators of cereal aphids. *Annual Report of Rothamsted Experimental Station for 1977*, p. 101.

Ge, F. and Chen, C.M. (1989). Laboratory and field studies on the predation of *Nilaparvata lugens* (Hom: Delphacidae) by *Theridion octomaculatum* (Aranae: Theridiidae). *Chinese Journal of Biological Control*, **5**, 84–88.

Grace, B. and Shipp, J.L. (1988). A laboratory technique for examining the flight activity of insects under controlled conditions. *International Journal of Biometeorology*, **32**, 65–69.

Greenslade, P.J.M. (1964). Pitfall trapping as a method for studying populations of Carabidae (Coleoptera). *Journal of Animal Ecology* **33**, 301–310.

Griffiths, E. (1983). *The Feeding Ecology of the Carabid Beetle* Agonum dorsale *in Cereal Crops*. PhD thesis, University of Southampton.

Hag Ahmed, S.E.M.K. (1990). *Biological Control of Glasshouse* Myzus persicae (Sulzer) *using* Aphidius matricariae *Haliday*. PhD thesis, University of London.

Halsall, N.B. and Wratten, S.D. (1988a). The efficiency of pitfall trapping for polyphagous predatory Carabidae. *Ecological Entomology*, **13**, 293–299.

Halsall, N.B. and Wratten, S.D. (1988b). Video recordings of aphid predation in a wheat crop. *Aspects of Applied Biology*, **17**, 277–280.

Hirano, M. (1987). Locomotor stimulant activity of fenpropathrin against the Carmine spider mite *Tetranychus cinnabarinus* (Boisduval). *Applied Entomology and Zoology*, **22**(4), 499–503.

Hulspas Jordan, P.M. and van Lenteren, J.C. (1978). The relationship between plant structure and parasitization efficiency of the wasp *Encarsia formosa* Gahan, Hymenoptera: Aphelinidae. *Mededelingen van de Faculteit Landbouwwetenschappen Rijksuniversiteit Gent*, **43**, 431–440.

Kennedy, J.S. (1978). The concepts of olfactory 'arrestment' and 'attraction'. *Physiological Entomology*, **3**, 91–98.

Kitching, R.L. and Zalucki, M.P. (1982). Component analysis and modelling of the movement process: analysis of simple tracks. *Researches of Population Ecology* **24**, 224–238.

Lewis, W.J. and Jones, R.L. (1971). Substance that stimulates host seeking by *Microplitis croceipes* (Hymenoptera: Braconidae), a parasite of *Heliothis* species. *Annals of the Entomological Society of America*, **64**, 471–473.

Luff, M.L. (1975). Some features influencing the efficiency of pitfall traps. *Oecologia (Berlin)* **19**, 345–357.

Mitchell, B. (1963). Ecology of two carabid beetles, *Bembidion lampros* (Herbst) and *Trechus quadristiatus* (Schrank). I. Life cycles and feeding behaviour. *Journal of Animal Ecology*, **32**, 289–299.

Mueller-Beilschmidt, D. and Hoy, M.A. (1987). Activity levels of genetically manipulated and wild strains of *Metaseiulus occidentalis* (Nesbitt) (Acarina: Phytoseiidae) compared as a method to assay quality. *Hilgardia* **55**(6), 1–23.

Nealis, V.G. (1986). Responses to host kairomones and foraging behaviour of the insect parasite *Cotesia rubecula* (Hymenoptera: Braconidae). *Canadian Journal of Zoology*, **64**, 2393–2398.

Obrtel, R. (1971). Number of pitfall traps in relation to the structure of the catch of soil surface Coleoptera. *Acta Entomologica Bohemoslavaca*, **68**, 300–309.

Penman, D.R. and Chapman, R.B. (1980). Effects of temperature and humidity on the locomotory activity of *Tetranychus urticae* (Acarina: Tetranychidae), *Typhlod-*

romus occidentalis and *Amblyseius fallacis* (Acarina: Phytoseiidae). *Acta Oecologica Applicata* **1**(3), 259–264.

Potts, G.R. and Vickerman G.P. (1974). Studies on the cereal ecosystem. *Advances in Ecological Research*, **8**, 107–197.

Riley, J.R., Smith, A.D. and Bettany, B.W. (1990). The use of video equipment to record in three dimensions the flight trajectories of *Heliothis armigera* and other moths at night. *Physiological Entomology*, **15**, 73–80.

Schaefer, G.W. and Bent, G.A. (1984). An infra-red remote sensing system for the active detection and automatic determination of insect flight trajectories (IRADIT). *Bulletin of Entomological Research*, **74**, 261–278.

Sunderland, K.D. (1975). The diet of some predatory arthropods in cereal crops. *Journal of Applied Ecology*, **12**, 507–515.

Sunderland, K.D. and Vickerman, G.P. (1980). Aphid feeding by some polyphagous predators in relation to aphid density in cereal fields. *Journal of Applied Ecology* **17**, 389–396.

Thorpe, K.W. and Dively, G.P. (1985). Effects of arena size on laboratory evaluations of the egg parasitoids *Trichogramma minutum, T. pretiosum*, and *T. exiguum* (Hym: Trichogrammatidae). *Environmental Entomology*, **14**, 762–767.

Treacy, M.F., Benedict, J.H. and Segers, J.C. (1984). *Effect of Smooth, Hirsute and Pilose Cottons on the Functional Responses of* Trichogramma pretiosum *and* Chrysopa rufilabris. Proceedings of the Beltwide Cotton Producers Research Conference, pp. 372–373.

Vinson, S.B., Harlan, D.P. and Hart, W.G. (1978). Response of the parasitoid *Microterys flavus* to the brown soft scale and its honeydew. *Environmental Entomology*, **7**, 874–878.

Waage, J.K. (1978). Arrestment responses of the parasitoid *Nemeritis canescens* to a contact chemical produced by *Plodia interpunctella. Physiological Entomology*, **3**, 135–146.

Wheater, C.P. (1989). Prey detection of some predatory Coleoptera (Carabidae and Staphylinidae). *Journal of Zoology*, **218**, 171–185.

4

Terrestrial molluscs

S.E.R Bailey

The leisured pace of snails and slugs and their nocturnal nature and preference for damp conditions make them tedious subjects for direct behavioural observations. The problems can be overcome by the use of time-lapse techniques. Initially, time-lapse cinephotography with synchronized flash was used (Newell, 1966; Gelperin, 1974), but this has been largely superseded by time-lapse video recorders and infra-red sensitive video cameras.

4.1 PRACTICAL ADVANTAGES AND PROBLEMS OF TIME-LAPSE

The merits and limitations of different techniques for studying snail and slug behaviour are compared in Table 4.1. Video equipment is adaptable, moderately priced and supplied 'ready to go', but the time-lapse recorder needs regular maintenance because it is in almost constant use. Image quality is lower than 16 mm cine film, but the recording can be checked immediately without a delay, while a whole film must be exposed and then processed (and only then, perhaps, is the exposure correctly adjusted). Video tapes are cheaper than film and can be reused. As with film, it is convenient to be able to keep written details of a trial on the same video tape as the results. Date and time are automatically added but, unlike cine, it is easy to superimpose the output from a computer data-logger. Unfortunately, the video apparatus and tapes attract thieves, and also video tapes can be accidentally erased, so that back-up copies are a wise precaution.

The major advantage over direct observation (apart from the researcher not having to stay up all night!) is that the recording can be replayed repeatedly to examine in detail the behaviour that precedes a particular

Video Techniques in Animal Ecology and Behaviour. Edited by Stephen D. Wratten.
Published in 1993 by Chapman & Hall, London. ISBN 0 412 46640 6

Table 4.1 The 'pros' and 'cons' of different observation techniques for slugs and snails

Attribute	Direct observation	Time-lapse cine	Time-lapse video
Field of view	Ideal. Can follow animals and view closely from any angle	View fixed, so animals fenced in, and arena two-dimensional	
Naturalness	Good, if no disturbance from observer	Impaired by above and by removing objects in field of view and by selecting a substrate which shows animals well	
Image quality	Excellent	Good	Reasonable
Recording long or complex behaviour	Poor: cannot record extra detail before an event	Excellent. Can replay repeatedly to analyse the details of behaviour preceding a specific event	
Simultaneous records from instruments	Difficult to tie instrument readings to specific events	Data superimposed by splitting field with mirror	Easy to superimpose computer output
Equipment costs	Nil. (BUT extreme boredom, loss of sleep and loss of time for other tasks)	High: pulse camera, flash sync unit, tripod and motion analysis projector	Moderate: infra-red camera and light, time-lapse VCR, tripod
Security	Risk to personal safety in night work	High risk of vandalism or theft of equipment	High risk of vandalism or theft of equipment, and of theft or accidental erasure of video tapes
Setting up	No problems	Correct exposure achieved by trial and error	Easy: instantly adjusted
Running costs	None	Costly and slow: wait until end of film, then wait for film to be developed.	Cheap and rapid: but recorder in constant use, so high maintenance cost
Analysis of behaviour	Low: observational cf. analytical	Depends on careful experimental design, but ability to record 'everything' may encourage poor experimental design	

event: the accurate recording of time and place every second is impossible for the direct observer. Slugs and snails spend most of their time inactive, and even replaying a time-lapse video recording gives hours of boredom, so many workers choose to record only the night hours. The ease with which video records 'everything' can encourage a sloppy attitude to experimentation, unlike other automated recording techniques which are designed from the outset to record a particular activity, encouraging careful experimental design.

The time-lapse system allows a more natural environment than the artificiality of Y-mazes (e.g. Chelazzi *et al.*, 1988), actographs (e.g. Lewis, 1969), or glass or polythene substrates on which mucus trails are traced (e.g. Tabor, 1987), but the camera has a fixed field of view, so if individuals are to be followed they must be artificially confined within an arena. The arena must be large enough to allow the animals to display their natural behaviour, but small enough so that the behaviour can be clearly seen. The Roman snail *Helix pomatia* does make seasonal 'migrations' between summer foraging and winter hibernating areas (Pollard, 1975). Fortunately, the species of greatest interest are large and most of their activities take place in a small area: they are the edible snails *Helix* and *Achatina*, the large limacid slugs *Limax* and *Limacus*, and pest species of slugs – *Deroceras*, the smaller *Arion* spp., and the keeled slugs *Tandonia* (= *Milax*). To get everything in focus, the arena is kept two-dimensional, and tall vegetation and other objects which might obscure the view are removed. Apart from the arena walls, the animals have few vertical surfaces and this may disturb their normal behaviour (e.g. *Helix aspersa* often mates on vertical surfaces). Having taken precautions to get a good view of the subjects, their most interesting actions are still almost certain to occur underneath the date and time register, which is usually electronically superimposed on the picture.

Confinement presents problems, not least how to confine these great escape artists which can scale the smoothest or roughest walls with equal ease; slugs can also squeeze through holes three-quarters of the diameter of their extended bodies, and glass lids mist over in fluctuating temperatures, so some ingenious fences have been designed. When similar individuals are kept in the same arena and come into contact, they may be difficult to distinguish when they separate.

4.2 RECORDING EQUIPMENT AND TECHNIQUES

4.2.1 Arena design

Size

Cine films have used large arenas: Newell (1966) filmed *Deroceras re-ticulatum* in an indoor arena 91 × 68 cm, and Gelperin (1974) filmed *Limax maximus* in an arena 1.7 × 1.5 m. Bailey (1989b) used an indoor arena 1.53 × 1.04 m to film distinctively painted *Helix aspersa*, and Cook (1980) filmed unmarked *Limacus pseudoflavus* in 1 m² of a flagged courtyard from a pulse camera mounted on the side of a house. Some video arenas have approached that scale: Cook (1979) video recorded *Limacus pseudoflavus* in an arena 1.15 × 0.58 m covered with ceramic tiles, and Bailey (reported by Wareing, 1986) recorded *Deroceras reticulatum* in an arena 97 × 97 cm. Most video recordings have been from smaller arenas, e.g. Howling (1991) used a soil-filled arena 57 × 36 cm indoors and similar areas outdoors were used by Howling and Port (1989) and Young and Port (1989). Young and Port used two markers 30 cm apart to indicate scale when a boundary fence was not used. North (1990), Munden and Bailey (1989), Bailey (1989a,b) and Bailey and Wedgwood (1991) used 30 × 30 cm arenas – wooden boxes filled with soil and a salt-impregnated strip of blotting paper stapled to the edges (Fig. 4.1). The advantage of these is that several can be used, so that slugs can 'settle in' for 2 days before the recording, and if needed two arenas can be videoed side by side.

Figure 4.1 Four wooden arenas, 30 × 30 cm, filled with soil and edged with dried salt-impregnated blotting paper, for simultaneous video recording. The slugs can shelter in depressions under the central tiles, and the surface is sprinkled with silver sand to show up the slugs (*Tandonia budapestensis*) well. Potato slices are placed in two arenas.

Clarity

Although Howling (1991) used 10 slugs of the same species in an arena, any more than three increases the chances of interactions, and the confusion of following a particular individual. The problem may be resolved by choosing individuals with differing pigmentation or of different size. Different genera can be distinguished fairly easily with practice. Clarity can be improved by choosing a 'top-dressing' of a substrate, e.g. silver sand or powdery soil, which best shows up the details of the specimens (this, however, makes the habitat artificial, and most slug workers have taken care to reproduce the natural features of a wheat field in their indoor arenas).

Slugs and snails cannot be recorded in dense vegetation, and Howling and Port (1989), to record the activity of natural populations of slugs in a grass/clover plot, uprooted the vegetation in the recorded area. This necessary alteration of the habitat unfortunately rules out some experiments it would be interesting to do, such as studying the competing attractiveness of slug pellets and crop plants (other than seedlings). Bailey and Wedgwood (1991) sank glass tubes part-way into the sand: these were adopted as daytime shelters (possibly because they provided thigmotactic stimuli), and the movements within the shelters could be seen. Some workers provide slugs with soil crevices (Newell, 1966) or opaque tiles (North, 1990), but shelters are seldom provided for snails.

Fences

Young and Port (1989) videoed slug movements into and out of an unfenced area of clover/grassland, but Howling and Port (1989) inserted a zinc barrier in the ground to a depth of 15 cm to prevent slugs wandering out of or into the area. Snail farmers use copper strips to deter escape (Moens *et al.*, 1986). The animals can be discouraged from exploring the sides of the arena by keeping the edges covered by dry loose sand, but Gelperin (1974) used a 2 cm strip of sodium chloride crystals, and North (1990) and Bailey and Wedgwood (1991) impregnated blotting paper with a saturated solution of sodium chloride and stapled it when dry to the sides of indoor wooden arenas.

Howling (1991) used an electric fence made of two parallel stainless steel wires of 0.25 mm diameter, 5 mm apart with a potential of 9 V between them. A sturdier weatherproof design (Fig. 4.2) uses the aluminium side of the arena as one conductor, turned inwards at the top to form a rain shelter. A second strip of aluminium underneath the shelter forms the second conductor, and is held away from the side by nylon bolts and screws. The 9 V battery is connected across the two conductors through a 10 KΩ resistor, to reduce excessive production of slime when slugs bridge the connector. Electric fences are not escape-proof: one snail can crawl over the

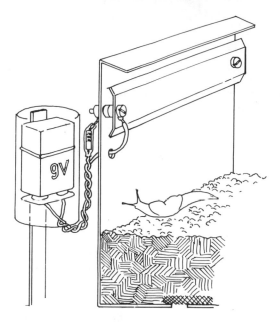

Figure 4.2 Section through edge of aluminium arena, showing electric fence.

shell of a second snail resting beside one conductor. Munden and Bailey (unpublished results) used a nylon net buried 50 cm deep and rising 80 cm high in a potato plot (Fig. 4.3): two tinned-copper wires were sewn onto the top of the net and connected to a 9 V battery, but did not entirely prevent *Arion rufus* climbing over in wet weather.

4.2.2 Environmental conditions

Indoor trials are usually conducted in a controlled-environment room, with controlled steady or cycling temperatures, cycling lighting, and high humidity maintained by mist sprayers – a surface film of water on soil is essential (Young and Port, 1991). Outdoors, temperatures at, above and below the soil surface, windspeed and humidity, rainfall, and solar radiation or light intensity are often recorded on to a data logger, but surface soil moisture is measured daily from a core dried at 90°C.

4.2.3 Lighting

Cook (1979) and Blanc *et al.*, (1989) used a red lamp for illumination in the dark phase, since pulmonates respond only weakly to long wavelengths (Kerkut and Walker, 1975). More commonly, lighting is provided by a 25 W tungsten bulb behind Wratten 87C gelatin filters, which provide light only

Figure 4.3 Panasonic camera inside aluminium shield and wrapped in polythene, supported on Benbo tripod, to record slug activity inside a netted area of an overgrown potato plot, in December. Infra-red lighting is from the outdoor spotlight through four gelatin filters sandwiched in glass.

from the near infra-red (wavelengths of 750 nm and longer). One overhead or two lateral lights are used at about 1 m from the ground, to avoid shadows. To simulate daylight, two 40 W fluorescent lights fixed 1 m above the arena will maintain a nocturnal rhythm, but Howling and Port (1989) used a 300 W quartz-halogen lamp 1.6 m above the soil surface, to give an intensity equivalent to an overcast day.

4.2.4 Studying underground activity

Munden *et al.* (in preparation) recorded the underground activity of slugs by filming through a vertical glass wall (Fig. 4.4). The slugs were confined to a 1 cm strip behind the glass by a nylon mesh sewn to a rigid wire frame and tucked down behind the top of the glass front to form an escape-proof seal but still let in light and air. Slugs would not burrow in loose loam, so an artifical system of channels was made between slabs of clay wired in position, and coated with soil particles. Extra channels were made, because in the field the slug would be moving in three dimensions to find additional channels. Various temperature gradients were established by building the soil box 45 × 45 cm wide and 30 cm deep, over a bed of sand with a cooling radiator in the base, and packing expanded polystyrene around the sides. Water was sprayed on to the entire surface, and could drain into the sand box beneath. The whole apparatus was placed in an unheated room, and light and heat provided by a 100 W photoflood lamp over the soil surface.

To video the side of the soil in darkness, a conical hood of black card was fitted between the glass and the camera lens; 25 W tungsten bulbs behind Wratten 87C infra-red filters were mounted on both sides of the cone, to light the soil without reflections. Black tape along the top of the glass, and a strip of black card behind the mesh at the soil surface, prevented light falling on the camera lens and shutting down the auto-iris.

Temperatures were recorded at the sides and centre of the profile at depths of 0, 5, 10, 15 and 20 cm using thermistor probes, and moisture levels

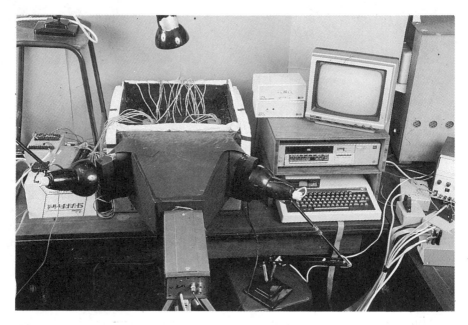

Figure 4.4 Apparatus for time-lapse video recording of the underground activities of slugs behind a glass screen, and superimposing records of temperatures, moisture levels and feeding. **a** Soil box in centre surrounded by expanded polystyrene with, in front, camera hood (with camera and with two lateral infra-red windows) covering glass front. Left of soil box are three probe amplifiers, and probe leads enter box from above. Right of box are BEEB-LOCK and monitor on top shelf, time-lapse VCR on centre shelf, and BBC B microcomputer beneath. Far right are multiplexers and signal conditioner and (not visible) a printer. **b** Details of connections, and monitor display.

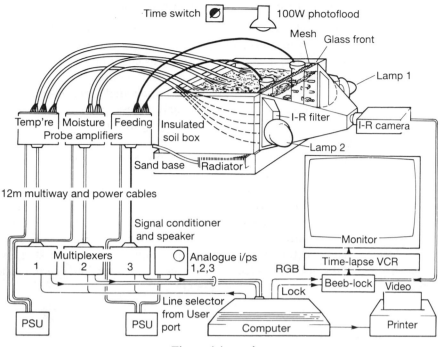

Figure 4.4 *contd.*

at 0, 6, 12 and 18 cm depths were recorded using silver wires embedded in plaster of Paris blocks. Ten feeding probes consisting of 5 mm diameter pellets of barley flour fixed on subminiature microphones were mounted beside the moisture detectors, at points congruent with the passages in the soil profile. The outputs from all probes were superimposed on the video recording.

4.2.5 Tripods

The video camera can be attached to a gantry, which also supports the lights. Outdoors, the superstructure must be uncluttered and slim to avoid altering the weather at ground level, for instance by causing large drips, and the camera needs a waterproof casing. If the video camera is used with a wide-angle lens, the legs of a standard tripod often enter the field of view. The Benbo 1 tripod has a central column which can be turned horizontally, overcoming this problem (see Fig. 4.3). The Benbo is steady enough to support a heavy video camera in a gale, but rainwater can enter the bottom sections of the legs unless the joints are covered with polythene.

4.2.6 Camera, lens and time-lapse video recorders

The most widely used video camera has been the Panasonic type WV-1850.IR fitted with an extended red Newvicon tube, sensitive to very low light levels (0.01 $\mu mol/m^2/s$), and an auto-iris lens enabling large changes in light intensity to occur between day and night without loss of picture quality. To film a large arena from above a wide-angle lens is used: this permits a lower ceiling height indoors, and a less obtrusive assembly outdoors. If mirrors are used to increase working distance they need to be large, and may mist up.

Recordings have been made without time-lapse equipment by switching a video recorder on for 6 s every minute (e.g. Cook, 1979), but the Panasonic AG6010 is a relatively inexpensive time-lapse video recorder used by many workers. When recording at 1/24th normal speed, a 3 h tape lasts for 3 days. The video tapes can be replayed on conventional recorders. The Panasonic 8050 has greater resolution, records at a wider choice of speeds, and has a less intrusive date and time register. High humidity is required for slug and snail activity, but causes the video tape to stick to the recording drum, so the recorder must be housed away from the arena.

4.2.7 Superimposed computer screens

Environmental parameters and other data, e.g. from electronic feeding probes, can be analysed on a microcomputer and superimposed on the video display. Munden *et al.* (in preparation) superimposed the display from a BBC model B computer on to the video screen using a VEL BL2K BEEB-LOCK (Video-Electronics Ltd, Manchester). The video recorder's date and time display was switched off and replaced by a computer display on a part of the screen where it did not obscure the video. If a lot of data are displayed, which would obscure the video screen, the data can be periodically blanked out by a program command to print 'black on black', and switched on again by a command to print 'white on black'.

For the complex information collected in the underground arena, the computer provides the binary code from the user port to select the line in three multiplexers dealing with temperature, moisture and feeding probes, as well as receiving the input from the peripheral amplifiers and buffers. The computer program has two parts: 'scan' and 'meal'. In 'scan', the program continuously scans each feeding probe to detect a bite, except once a minute when it scans the temperature and moisture probes. Temperatures and moisture levels are displayed on the monitor screen at positions corresponding to the position of the probes in the soil. When a slug starts to feed, the program switches to 'meal', locking on to the feeding probe where the bite has been detected. Meal characteristics (number of bites, bite interval etc.) are displayed along the top of the screen. After the meal, the program reverts to 'scan'.

4.2.8 Analytical tools: digitizers

Analyses fall into two major categories: temporal and spatial. Temporal analysis (duration and frequency of various activities and transitions between them, etc.) requires no further hardware because each frame contains a date and time registration. Simple spatial analyses of distances or speed also may need no further equipment. Lorvelec (1990) and Lorvelec *et al.* (1991) divided a glass snail arena $25 \times 25 \times 10$ cm into 5×5 cm squares to register position in all three dimensions; using a digitizer would have required a complicated program, because of the distorted image of the walls. Lorvelec *et al.* (1991) discuss the subsequent resolution of the problem of equating movement (measured as distance) and feeding (measured as time spent). However, many two-dimensional analyses (distances moved, angles turned, areas covered etc.) are speeded up by using a digitizer connected to a microcomputer, e.g. a Grafbar 7 sonic digitizer (Fig. 4.5). Although the tracks could be digitized straight from the video monitor screen, in practice the tracks are first transcribed on to acetate sheets fixed to the monitor screen. Times are

Figure 4.5 Using a Graf-Bar 7 sonic digitizer to analyse tracks transferred from a video monitor on to acetate sheet. The digitizer is the rectangular wall-mounted box and the digitizer pen is pressed against each point to be digitized. The computer display reproduces the track and records scaled distances along the bottom. Along the top of the monitor screen and beneath the digitizer is the menu for selecting distance, angle, area etc.

indicated on the tracks by a mark every minute, and by numbering each pause
on the track with a corresponding record of the time of each numbered pause
written in the margin. Figure 4.6a shows the kinds of features of a track
commonly analysed, and Fig. 4.6b shows a specimen acetate record.

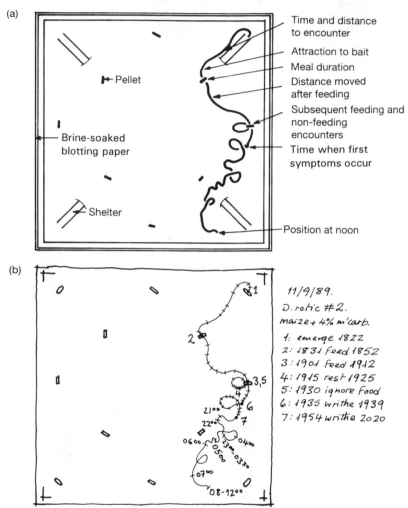

Figure 4.6 (a) Features of tracks recorded. (b) Specimen acetate record.

4.3 USES OF VIDEO TECHNOLOGY

4.3.1 Simple measurements

Video recording is so simple and rapid that freeze frames are used instead of
still photos to measure the foot during crawling, or the delicate shells of

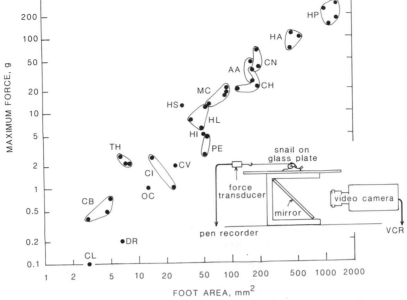

Figure 4.7 Use of a normal video camera with macro facility, and ordinary video recorder with freeze-frame facility to measure foot area of 17 different species of land snail to study relationship between foot area and maximum pull (from Bailey, 1992a). AA: *Arianta arbustorum*; CB: *Clausilia bidentata*; CH: *Cepaea hortensis*; CI: *Candidula intersecta*; CL: *Cochlicopa lubrica*; CN: *Cepaea nemoralis*; CV: *Cernuella virgata*; DR: *Discus rotundatus*; HA: *Helix aspersa*; HL: *Helicigona lapidica*; HI: *Helicella itala*; HP: *Helix pomatia*; HS: *Hygromia striolata*; MC: *Monacha cantiana*; OC: *Oxychilus cellarius*; PE: *Pomatias elegans*; TH: *Trichia hispida*.

juveniles (Fig. 4.7). Video recordings have also been used to measure crawling speed and mucus viscosity (by measuring the rate of descent on a mucus thread (Henderson, pers. comm.)). The system can also be used to check equipment performance, such as the efficiency of traps.

4.3.2 Circadian rhythms and time budgets

Circadian rhythms are usually studied in isolated animals, but time-lapse infra-red video recording enables the phenomenon to be followed in individual members of groups of slugs (Ford, 1986) and snails (Lorvelec *et al.*, 1991). Time budgets are also easily constructed from video recordings (e.g. Cook, 1979; Bailey, 1989b; Bailey and Wedgwood, 1991). The activity records usually show a high peak of crawling after dusk, and a lesser peak around dawn (Fig. 4.8). Blanc *et al.* (1989) compared the activity rhythms of

Terrestrial molluscs

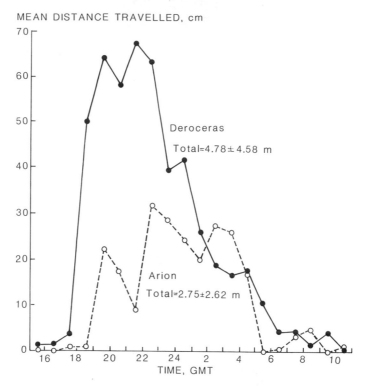

Figure 4.8 Mean distance moved each hour by *Deroceras reticulatum* and *Arion distinctus* (from Bailey and Wedgwood, 1991).

two helicid species when kept in single-species groups and in a combined-species group: both species are commercially important, and co-rearing might permit high stocking densities without depression of growth. They found that *Helix pomatia* were more active at dawn and *H. aspersa* more active at dusk, and there was no noticeable modification when the two species were in the same cage.

4.3.3 Effects of weather and hunger on activity

Young and Port (1991) showed that *Deroceras reticulatum* was active for longer on the surface of soils with a high moisture content in the top 2.5 cm. The effects of temperature on slugs on travel distance and feeding time were examined briefly by Munden and Bailey (1989). Brief video studies of the effect of hunger on slug and snail activity were made by Bailey (1989a,b). Bailey (1989b) found that adult snails spent longer feeding than juveniles, and that snails moved faster towards food than away from it.

Young and Port (1989) recorded slug movements on a grass/clover plot to produce a predictive model of slug movements under different weather conditions. Activity was measured as the total length of tracks of all slugs that appeared in the field of view. The microclimatic variables that showed the greatest correlation with activity were built into a 'limit'-type model, such that only when all environmental variables favoured activity was high activity predicted. The predicted high-activity nights were then tested against the numbers of slugs collected from bran-baited slug traps on an adjacent field on the same nights. The model correctly predicted 17 of 20 high-activity nights. Activity models on an hourly basis have been developed by Ford (1986), using cine film of *Limacus pseudoflavus*, and by Rollo (1982) using direct observation of *Limax maximus*.

4.3.4 Trail following

Trail following occurs in all three subclasses of gastropods, but the studies usually employ a laborious technique of 'developing' trails with a fine mist of water vapour or dipping in a suspension of powdered talc or chalk and washing (e.g. Tabor, 1987). An often-quoted measure of trail following is Townsend's index of coincidence, i.e. the length of trail followed divided by the square root of the product of the lengths of marker and follower trails (Townsend, 1974). However, Cook (1977) points out that the coincidence index does not distinguish between number of instances of following and distances followed, and it is dependent upon the distances moved by the animals. Trail following is often seen in video recordings of various slug species, but especially in *Deroceras reticulatum*, and sometimes frustrates the purpose of the experiment – as when a slug follows its own trail in a closed loop for most of the night, or tracks another individual and crawls over but ignores food pellets which are the subject of the investigation. Wareing (1986) reported several instances of trail following in *Deroceras reticulatum*, usually in the direction that the trail was laid, sometimes over long distances, and often ending with copulation or attempted copulation (Fig. 4.9).

4.3.5 Homing

Many of the larger or longer-lived terrestrial gastropods home to a daytime roost. Newell (1966) observed the return of *Deroceras reticulatum* to the same soil crevice after foraging, and Gelperin (1974), using time-lapse cine, found that *Limax maximus* would respond to home odour from 1 m away. Cook (1979, 1980) studied the roles of trail following and distant odours in homing in *Limacus pseudoflavus*. He postulated that a slug's best strategy while foraging would be to stay downwind of home and thus maintain olfactory contact, and use trail following to home when distance olfaction is disrupted by variable winds.

80 *Terrestrial molluscs*

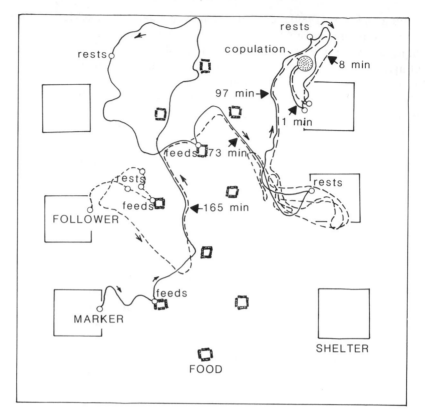

Figure 4.9 Trail following in *Deroceras reticulatum* recorded in an arena 97 × 97 cm, ending in copulation. The times along the tracks show how the delays of the follower slug shorten until it catches up with the marker slug.

Lorvelec (1990) and Bailey (1989b) used time-lapse video and cine to study homing in *Helix aspersa* in two very different arenas: Lorvelec used a small glass arena, and Bailey used a large arena of soil containing two piles of rocks and wood. Bailey concluded that return was to the general area, not – as in limpets – to the precise position, and routes between roost site and food site became more direct with time (Fig. 4.10). Tabor (1987) used a simple index of directness (Hamilton, 1977) to show that exploratory behaviour decreased in the slug *Limacus pseudoflavus*, and it moved with greater accuracy between shelters and food on successive nights in an arena. However, they continued to make excursions, probably to find additional shelter sites. Even in Lorvelec's simple environment, *Helix aspersa* made periodic returns eight times more frequently than would be expected by chance, and microclimatic factors could be ruled out. In addition, Lorvelec saw that homing was not related to aggregation but depended solely on the individual. In both snail studies, individual animals

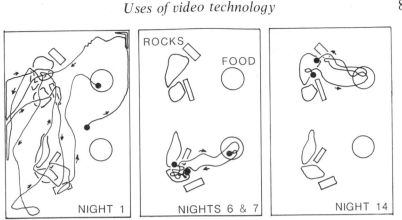

Figure 4.10 Changes in the tracks of *Helix aspersa* over a period of days in the same arena.

were found to use several roost sites, which they shared with other individuals.

4.3.6 Dispersal

Slugs and snails have been used as models for random walks. Hunter and Symonds (1970) considered the ideal distribution density of slug pellets, based on a random movement pattern. Bailey (1989a) used turn angles along every 2 cm of track in a video arena to produce a model of dispersion. Increasing track lengths, increasing the concentration of angles in the forward direction, or a decrease in circling (successive turns in the same direction) only slightly increase the mean distance of displacement, but they greatly increase dispersion. Most tracks show a distribution of turn angles similar to a theoretical von Mise's distribution with a K of 4 (Batschelet, 1981), but with rather more large turns; the mean angle is usually within 5° of zero, and the sequence of left and right turns is almost random.

4.3.7 Foraging

Although Duval (1970) found no evidence of turning as slugs passed close by pelleted food, Port and Hogan (1985) produced recordings of slugs 'working their way' along a row of drilled cereal grains, and Airey *et al.* (1989) observed highly significant increases in slug activity within a 60 mm radius zone around samples of lettuce and dandelion volatiles, compared to a water control. Using Kareiva and Shigesada's (1983) dispersion model, Howling (1991) compared actual slug foraging tracks with a predicted path calculated by assuming that the forager moves randomly. Predicted and observed displacements agreed when no food was available, but when certain bait pellets were provided the slugs showed greater displacement than predicted. Bailey and Wedgwood (1991) found that the rate at which slugs encountered food items was proportional to the length of their tracks, but slugs which would feed found food five times as frequently as non-feeding slugs. This might have been achieved by moving the head from side to side to survey a broader track, by circling back to food soon after leaving it, or by olfaction, indicated by sharply turning on to food from about one body length away. Howling (1991) showed that slugs moved further than predicted by a random displacement model at the start of a foraging trip, but then turned more sharply, perhaps seeking a crevice or finding food.

4.3.8 Feeding

Slugs and snails feed several times through the night and, when feeding on pelleted food, slugs almost always accept the first food pellet encountered, rejecting an increasing proportion of contacted items later (Bailey, 1989a). Over half of each night's food was taken within an hour of starting to feed, especially if the slugs had been deprived of food. However, slugs feeding on germinating wheat grains rejected more than half of the first grains found (Bailey and Wedgwood, 1991), and North (1990) showed that on soft foods, such as carrot, the second meal may last as long as the first. North also noted that there were interspecific differences: *Milax* was a more selective feeder than *Arion distinctus*, possibly because of its larger foraging area, both above and below ground.

When slugs were given a choice of a more and a less palatable food (maize and pea flour pellets, respectively), they encountered each type in proportion to their abundance, but ate more from the maize pellets unless the pea pellets were seven times more common (Bailey, 1989a). Bailey (1992b) recorded bites on a pellet electronically and superimposed them on a video recording, to show that fed slugs (but not starved ones) ate less overall if they fed first on the less palatable pea flour (Fig. 4.11). Slugs would often take a few 'sample' bites on a pellet before subsequently ignoring it.

In general, video recordings produce less precise records of meals than electronic recordings of feeding (the 'acoustic pellet' technique), because the video recording fails to show whether the slug feeds all the time that it is in contact with a food item (Bailey and Wedgwood, 1991). However, video is needed to show that, after a meal, a slug may rest in contact with the food and then resume feeding, or may leave and return to resume feeding within a few minutes. Of course, softer foods such as leaves can be seen to disappear, and Bailey (1989b) noted that *Helix aspersa* spends a good deal of time on the food leaf without feeding, especially at the end of a meal, perhaps indicating increasing selectivity.

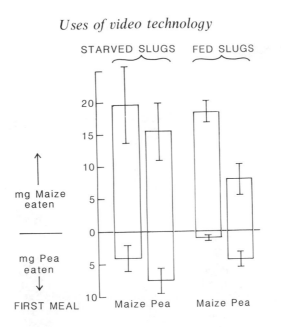

Figure 4.11 The influences of hunger and of type of pellet eaten first on food consumption by *Deroceras reticulatum*, when given a choice of a common but less palatable food (pea flour pellets) and a less common but preferred food (maize flour pellets) (from Bailey, 1992b).

4.3.9 Poisoning and recovery

Much of the attention given to slug foraging is directed at their response to molluscicidal baits (e.g. Howling and Port, 1989; Bailey and Wedgwood, 1991). The addition of metaldehyde or, to a lesser extent, the carbamate methiocarb, reduces meal length dramatically compared to non-molluscicidal foods, and it seems difficult to improve on the palatability of cereal-based baits.

The use of acoustic-pellet recordings superimposed on a real-time macro video recording of slugs feeding on different types of pellet shows the effect of metaldehyde baits in producing misdirected rasping movements at the end of the foreshortened meal (Fig. 4.12). Howling and Port (1989) found that baits containing metaldehyde produced a toxic effect on *Deroceras reticulatum* much more rapidly than those containing methiocarb only. After stopping feeding, slugs remained in contact with the metaldehyde baits for a long time; after a methiocarb meal they moved a greater distance, often going underground. Similar results were obtained by Bailey and Wedgwood (1991), who saw locomotor impairment of methiocarb-fed *Deroceras* only about 3 hours after the meal, and 58% found shelter (although *Arion distinctus* became immobilized much more rapidly). In contrast, after a metaldehyde meal both species moved little, seldom encountered new

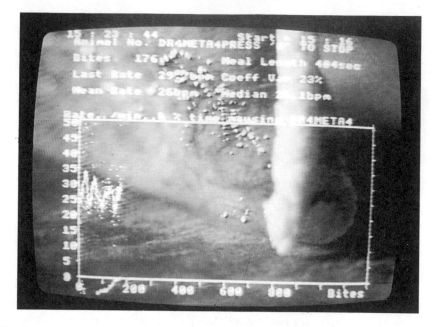

Figure 4.12 Normal-speed video recording made with macro-lens camera, of a slug at the end of a meal on a pellet containing the molluscicide metaldehyde, showing the misdirected biting (jaw and radula everted to the side of the pellet). A computer analysis of the meal, using a subminiature microphone insert to record vibrations, is electronically superimposed on the recording and records the 176 bites before the meal ended.

pellets, seldom fed again, and only 13% found shelter. After 45 minutes, metaldehyde-fed *Deroceras* were writhing or producing a thick pad of mucus.

4.3.10 The underground lives of slugs

Munden *et al.* (in preparation) showed that over half of the colony of *Tandonia* (= *Milax*) *budapestensis* studied settled more than 15 cm below the surface, but *Deroceras reticulatum* were equally distributed between the surface and 15 cm (Fig. 4.13). There was only a slight upward shift in distribution at night, and slugs remained underground unless the soil was watered. They showed no tendency to change position to follow the strongly fluctuating daily temperatures, but did maintain a strong nocturnal pattern of activity and feeding. Meals were infrequent, but when an animal settled on a pellet several meals were taken in succession, and the animal stayed near the pellet in between meals. The duration of these batches of meals varied from 7 to 49 minutes.

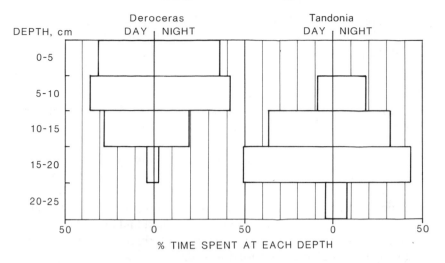

Figure 4.13 Vertical distribution of *Deroceras reticulatum* and *Tandonia budapestensis*, by day and night, recorded by infra-red time-lapse video through the glass wall of an artificial soil profile.

4.3.11 Aggression and courtship

When several animals are kept together, only cine or video records can follow the complex interactions. Courtship and mating in snails was studied by Beaumont (1988), and Bailey (1989b) recorded the incidence of courtship in *Helix aspersa*. North (1990), in an investigation of competition between sympatric species of slugs, found interspecific differences in the behaviour of five slug species: most showed increased time crawling in the presence of conspecifics, but *Deroceras caruanae* became less active. She investigated the frequency of interactions within species and between pairs of species of slugs which commonly occur together (Table 4.2). *Deroceras caruanae* frequently

Table 4.2 Frequency of aggressive (a) and non-aggressive (non-a) encounters between three conspecifics or two pairs of slugs of different species, in a 30×30 cm arena over 3 nights. (After North, 1990.)

	Intraspecific encounters	*Interspecific encounters initiated*	*Interspecific encounters received*
	a: non-a	*a: non-a*	*a: non-a*
Deroceras caruanae	24:1	18:4	2:4
Deroceras reticulatum	5:4	8:5	6:5
Tandonia budapestensis	4:3	0:4	5:3
Arion subfuscus	1:2	0:3	7:3
Arion distinctus	0:0	0:5	6:6

shows aggression both to conspecifics and other species – usually biting the tail. When attacked they may retaliate, or flee while lifting the tip of the tail clear of the ground and waving it from side to side.

4.4 CONCLUSIONS

Time-lapse photography brought a new dimension to our perception of many ecological processes such as growth and decay, and proved eminently suitable for studying the movement patterns of slow-moving molluscs over one or several nights. Its principal advantages are that it allows you to reverse time in order to pay attention to what was happening before an event occurred, and that it allows individuals to be followed in complex interactions. With the advent of video recording, time-lapse techniques became much easier and quicker to use, and should find increasing uses. The electronic basis of video imaging also enables information from data-loggers to be superimposed on to the visual record.

What of the future? Technological progress is already leading to automatic digitization of the images, and their storage on data-loggers for easier processing. But is there also a conceptual advance required? Current systems are limited by the fixed view of the camera: if it covers all the area the animals need, it cannot resolve in detail what they are doing, and it cannot see under vegetation. Electronic zooming exists, but a human observer seems to be the ideal monitoring system, equipped with pan, tilt and zoom eyes, and able to lock on to and follow individuals over rough ground. The principal drawback to developing an equally flexible system now is that we could not afford to buy it.

REFERENCES

Airey, W.J., Henderson, I.F., Pickett, J.A., *et al.* (1989). *Novel Chemical Approaches to Mollusc Control.* British Crop Protection Council Symposium 41 (Slugs and snails in world agriculture, (ed. I. Henderson)), 301–307.

Bailey, S.E.R. (1989a). Foraging behaviour of terrestrial gastropods: integrating field and laboratory studies. *Journal of Molluscan Studies*, **55**, 263–272.

Bailey, S.E.R. (1989b). Daily cycles of feeding and locomotion in *Helix aspersa Haliotis*, **19**, 23–31.

Bailey, S.E.R. (1992a). Speed and strength in land snails. *Abstracts of 11th International Malacological Congress, Sienna* (eds F. Giusti and G. Manganelli), 386–387.

Bailey, S.E.R. (1992b). Foraging behaviour of terrestrial gastropods: effects of changing level of food arousal on consumption. *Proceedings of 10th International Malacological Congress, Tübingen, 1989*, 421–424.

Bailey S.E.R. and Wedgwood, M.A. (1991). Complementary video and acoustic recordings of foraging by two pest species of slugs on non-toxic and molluscicidal baits. *Annals of Applied Biology*, **119**, 163–176.

Batschelet, E. (1981). *Circular Statistics of Biology*, Academic Press, London.

Beaumont, M.A. (1988). *A Study of Reproductive Interference Between Closely Related Species.* Unpublished PhD thesis, University of Nottingham.

Blanc, A., Buisson, B. and Pupier, R. (1989). Evolution en laboratoire du rythme spécifique d'activité de deux mollusques gastéropodes (*Helix pomatia* L. et *Helix aspersa* Muller) en situation de cohabitation sous différentes photopériodes. *Haliotis*, **19**, 11–21.

Chelazzi, G., Le Voci, G., Parpagnoli, D. (1988). Relative importance of airborne odours and trails in the group homing of *Limacus flavus* (Linnaeus) (Gastropoda: Pulmonata). *Journal of Molluscan Studies*, **54**, 173–180.

Cook, A. (1977). Mucus trail following by the slug *Limax grossui* LUPU. *Animal Behaviour*, **25**, 774–781.

Cook, A. (1979). Homing by the slug *Limax pseudoflavus*. *Animal Behaviour*, **27**, 545–552.

Cook, A. (1980). Field studies of homing in the pulmonate slug *Limax pseudoflavus* (Evans). *Journal of Molluscan Studies*, **46**, 100–105.

Duval, D.M. (1970). Some aspects of the behaviour of pest species of slugs. *Journal of Conchology*, **27**, 163–170.

Ford, D.G.J. (1986). Rhythmic activity of the pulmonate slug *Limax pseudoflavus* Evans. Unpublished DPhil Thesis, University of Ulster.

Gelperin, A. (1974). Olfactory basis of homing in the giant garden slug, *Limax maximus*. *Proceedings of the National Academy of Sciences USA*, **71**, 966–970.

Hamilton, P.V. (1977). Use of mucus trails in gastropod orientation studies. *Malacological Reviews*, **10**, 73–76.

Howling, G.G. (1991). Slug foraging behaviour: attraction to food items from a distance. *Annals of Applied Biology*, **119**, 147–153.

Howling, G.G. and Port, G. (1989). *Time-Lapse Video Assessment of Molluscicidal Baits*. British Crop Protection Council Symposium 41; Slugs and snails in world agriculture (ed. I. Henderson), 161–166.

Hunter, P.J. and Symonds, B.V. (1970). The distribution of bait pellets for slug control. *Annals of Applied Biology*, **65**, 1–7.

Kareiva, P.M. and Shigesada, N. (1983). Analysing insect behaviour as a correlated random walk. *Oecologia, Berlin*, **56**, 234–238.

Kerkut, G.A. and Walker, R.J. (1975). Nervous system, eye and statocyst, in *Pulmonates, Vol. 1* (eds V. Fretter and J. Peake), Academic Press, London, pp. 165–244.

Lewis, R.D. (1969). Studies on the locomotor activity of the slug *Arion ater* (Linnaeus) I. Humidity, temperature and light reactions. *Malacologia*, **7**, 307–312.

Lorvelec, O. (1990). Le retour au gite chez l'escargot *Helix aspersa*. Etude au laboratoire. *Biology of Behaviour*, **15**, 107–116.

Lorvelec, O., Blanc, A., Daguzan, J., Pupier, R. and Buisson, B. (1991). Etude des activités rythmiques circadiennes (locomotion et alimentation) d'une population Bretonne d'escargots *Helix aspersa* Muller en Laboratoire. *Bulletin de la Société Zoologique, France*, **116**(1), 15–25.

Moens, R., Gigot, J. and Vase J. (1986). Copper sheet, an effective barrier in the rearing of *Helix pomatia* (L). *Snail Farming Research* (Associazione Nazionale Elicicoltori) **1**, 23–26.

Munden. S., and Bailey, S.E.R. (1989). *The Effects of Environmental Factors on Slug Behaviour*. British Crop Protection Council Symposium 41; Slugs and snails in world agriculture (ed. I. Henderson), 263–272.

Newell, P.F. (1966). The nocturnal behaviour of slugs. *Medical and Biological Illustration*, **16**, 146–159.

North, M. (1990). *The Distribution of Terrestrial Slugs and Snails in the North West of England and a Behavioural Study of Five Commonly Occurring Slug Species*. Unpublished MSc thesis, University of Manchester.

Pollard, E. (1975). Aspects of the ecology of *Helix pomatia*. *Journal of Animal Ecology*, **44**, 305–329.

Port, G.M. and Hogan, J.M. (1985). *Foraging behaviour of the grey field slug* Deroceras reticulatum *(Muller) with respect to food and bait.* Unpublished proceedings of an AAB/SEB workshop on the use of video techniques in pure and applied entomology held at the University of Southampton, 7–8 January 1985.

Rollo, C.D. (1982). The regulation of activity in populations of the terrestrial slug *Limax maximus* (Gastropoda: Limacidae). *Researches on Population Ecology*, **24**, 1–32.

Tabor, A.M. (1987). *The Effects of Conspecific Mucus on the Behaviour of Some Gastropod Molluscs.* Unpublished PhD thesis, University of Manchester.

Townsend, C.R. (1974). Mucus trail following by the snail *Biomphalaria glabrata* (Say). *Animal Behaviour*, **22**, 170–177.

Wareing, D.R. (1986). Directional trail following in *Deroceras reticulatum*. *Journal of Molluscan Studies*, **52**, 256–258.

Young, A.G. and Port, G.R. (1989). *The Effect of Microclimate on Slug Activity in the Field.* British Crop Protection Council Symposium 41; Slugs and snails in world agriculture (ed. I. Henderson), 263–272.

Young, A.G. and Port, G.R. (1991). The influence of soil moisture content on the activity of *Deroceras reticulatum* (Muller). *Journal of Molluscan Studies*, **57**, 138–140.

5

Marine video

C.S. Wardle and C.D. Hall

5.1 INTRODUCTION

We are all familiar with the colourful underwater world from the ever-popular films of brightly coloured fish on coral reefs. This shallow and brightly lit zone has lent itself to successful filming even with cine systems. Scuba divers can hire or buy from a choice of modern self-contained automatic colour 'Camcorders' in well designed underwater housings, and there seem no limits even for the amateur. However, go just a little bit deeper in any part of the ocean and things get more difficult. The domestic cameras commonly available are designed to work in normal daylight ranges of light level, and even in the clearest water they become of little use below 30 m depth unless artificial lights are used. This chapter looks at how this may be overcome, asks what are the limits to observing with natural light, what effects will artificial light have and suggests the basis for being aware of and solving some of the problems of intrusion and avoiding disturbance of the natural behaviour of animals. The principles of using towed and self-powered vehicles for surveying larger areas and keeping up with moving targets like fishing gears both with and without Scuba equipment and the advantages of remote control are discussed. The bright colours of the popular underwater films are misleading, as most of this marine world is gloomy and lacking in colour and, more importantly, many of the animals when in natural light and surroundings are as near invisible as they can be.

5.1.1 What type of TV cameras are suitable?

The historical record of the use of television in the sea is not particularly useful other than to indicate that we are a great deal better off since 1974

Video Techniques in Animal Ecology and Behaviour. Edited by Stephen D. Wratten. Published in 1993 by Chapman & Hall, London. ISBN 0 412 46640 6

when the silicon-diode intensified target (SIT) camera tube was made available by RCA and first built into cameras for underwater use (Wardle, 1993). Attempts to use cameras before this time were not particularly productive, except in founding experimental approaches for deploying cameras in hostile situations. These attempts to use both photography and television to record information on various aspects of the marine environment have been well documented (Barnes, 1963; George *et al.*, 1985) since its origins in the late 1940s. Rapid advances in technology are still taking place, making it essential to keep up with current scientific and technical literature when planning a research programme. The following article briefly mentions examples of modern applications of video and TV equipment in underwater research, many of which have been developed by staff of the Marine Laboratory, Aberdeen, and used in fisheries laboratories worldwide.

Observation of towed fishing gear was one of the main challenges to scientific divers using photographic and direct observation techniques between 1952 and the early 1970s (Main and Sangster, 1978; Wardle, 1983, 1985, 1993). The availability of an underwater camera fitted with the SIT tube opened up large new areas of the sea where video recordings could be made in the low light conditions found in normal towed trawling conditions. The camera allowed filming of fast-moving scenes and gave bright clear pictures, whereas smearing, ghosting and comet-trailing of moving images was a problem of vidicon TV cameras. A recent comparison of the performance of camera tube types at different light intensities is given in Table 5.1. It shows that SIT and ISIT (intensified SIT) tubes stand out among the others, particularly at lower light levels. They generate bright enhanced contrast images at light levels as low as -3 and -4 log lux respectively. In comparison, the human dark-adapted eye is still useful at -4 log lux. The TV camera still has the potential to become more useful to the biologist. There are indications that cameras with up to 50 times the ISIT sensitivity but using CCD devices with intensifiers may soon be available. The ISIT camera has a small sensitivity advantage over the SIT that could be significant in some applications.

A camera fitted with an SIT tube is more expensive, but SIT tubes have a very important advantage in a wide range of circumstances in animal behaviour studies underwater, where daylight levels are inevitably reduced and a less sensitive camera would need artificial light which would interfere. There is an added advantage in more turbid water, where the use of a light source would result in deterioration of the TV picture due to backscatter. The light source, however well arranged, inevitably illuminates suspended particles, creating a veil over the picture. Cameras fitted with the SIT tube are used extensively around the world for fisheries observations because of these advantages.

Table 5.1 Comparison of the performance of TV camera pick-up tubes at different light intensities. The following conditions are assumed in all cases: (1) cameras are fitted with T2 lens; (2) scene reflectance of 50% (from Harris, 1980.)

Tube type	10^{-4}	10^{-3}	10^{-2}	10^{-1}	1	10	10^2	10^3	10^4	10^5
1. Plain vidicon (sulphide type)						Good, very high definition pictures by target control, marred by excessive 'memory' at lower light levels				N.D.* filter or iris needed to prevent 'blooming'
2. Newvicon Chalnicon Silicon Vidicon					Fading into amplifier noise	Good, high definition pictures with low lag			Fading into amplifier noise and vidicon defects	
3. SIT (16 mm) tube			Fading into photoelectron and amplifier noise	Good, very clear pictures			Very good pictures can be maintained with N.D.* filter on lens			
4. ISIT (16 mm) tube	Fading into photoelectron noise	Good clear pictures					Good clear pictures with N.D.* filter on lens			
Scene illum. lux	10^{-4}	10^{-3}	10^{-2}	10^{-1}	1	10	10^2	10^3	10^4	10^5
Viewing conditions	Starlight Overcast Clear		Moonlight $\frac{1}{4}$ Moon	Full		Twilight	Room Lighting		Daylight Overcast	Full Sun

*Neutral density. (**R**eproduced with permission from Harris, 1980.)

5.2 PROBLEMS SPECIAL TO UNDERWATER CAMERAS

5.2.1 The underwater light

In underwater conditions daylight intensity is attenuated as depth increases: the rate of change varies with wavelength and clarity of the water.
For instance, below 20 m in coastal waters various colours are filtered
from the light by the column of water. The resulting colour depends on
components of the water column and the distance. The result in general
is that the daylight illuminates objects and their backgrounds with
greenish light from above. Figure 5.1, from Mackay (1991), is based on
the spectral response of pure, clear water, illustrating one of the fundamental properties of water in that it does not transmit all the visible
wavelengths of light equally. Maximum transmission occurs at the green
end (about 450 nm) of the light spectrum, where attenuation is only of the
order of 2% (over a 1 m path). However, at the red end of the spectrum,
around 650 nm, transmission is reduced by about 73%. As the length of
the light path increases, the effect becomes more pronounced. Over a 4 m
path the transmission of green light is still above 90%, but red light is
reduced to about 30%. There are seasonal changes to water components
which themselves can affect the water colour and transparency. Water
masses move and form stratified layers, and there is always the potential
for large variations in these important qualities, even locally and from day
to day.

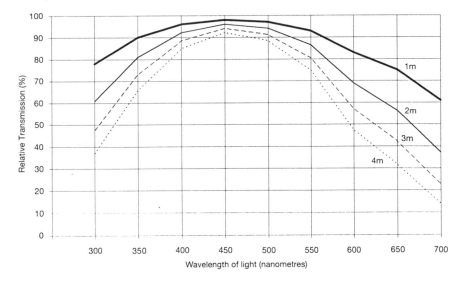

Figure 5.1 The relative spectral response of seawater across the visible light range,
at different viewing distances (reproduced with permission from Mackay, 1991).

A practical way of illustrating the effect of depth on colour is to mount a depth gauge and colour test card in front of a colour camera such as the scuba diver's Sony Handycam, lower it to 40 m and, once recovered, look at the video film. The red colours have gone in the first 5 m, by 15 m only green is seen, by 20 m the image contrast is very low, and by 30 or 35 m it is just too dark. It is worth bearing in mind that even when artificial light is used with colour cameras the colour will be modified as the range of viewing increases. Also be aware that the colour balance of colour cameras can be reset for different light source types to obtain a more natural daylight appearance.

SIT and ISIT monochrome cameras are, however, ideally suited for use in deeper marine conditions because the maximum sensitivity of the cathode of the intensifier occurs in the blue–green part of the spectrum, coinciding with the maximum light transmission characters of seawater. Notice in Fig. 5.2, (from Tyler and Smith, 1967), how at a depth of 100 m the energy level of

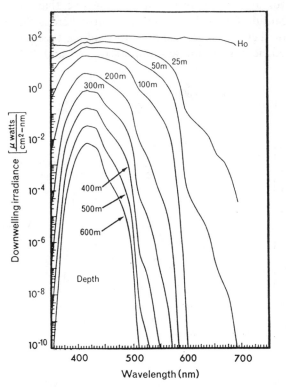

Figure 5.2 The downwelling irradiance at various depths in the exceedingly clear waters of Crater Lake, Oregon. These irradiances have been calculated from K values measured in the lake, and assume the lake is homogeneous. (From Tyler and Smith, 1967, by permission of Oxford University Press.)

green light of between 400 and 450 nm is about 11 orders of magnitude greater than red light.

5.2.2 Light level limits to cameras and animal behaviour

Figure 5.3 (from Cui *et al.*, 1991) shows the variation of light at the sea surface at latitude 60°, and indicates how seawater modifies the light intensity with depth and how different conditions of transparency might modify the rate of light level change with depth (Fig. 5.3b). Figure 5.3b may be lowered relative to Fig. 5.3a in order to consider the attenuation due to depth of any surface light level. On this diagram are also marked a number of thresholds, including the natural light limits of the various TV camera types. The reaction threshold for mackerel schooling and fish reactions in fishing gears is also shown, at about −6 log lux. In general, most vertebrates that have been studied sufficiently well have shown us that their eyes are several orders more sensitive than the best available cameras. A caution for the behavioural scientist here is that if their camera cannot see, or their light meter shows no reading, this does not mean that there is insufficient light for an animal to see and behave with visual reactions.

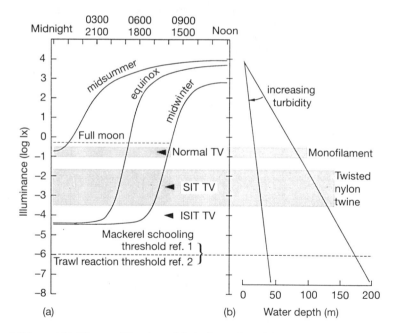

Figure 5.3 (a and b) The modification of light intensity with depth and the variation of surface light levels at latitude 60° north. The shaded areas are thresholds for mackerel reacting to the fishing twines indicated. The ISIT camera could be used to observe these thresholds, but not the SIT (modified from Cui *et al.*, 1991).

5.2.3 How to measure such low light levels

TV cameras and light meters stop working at about -4 log lux, at a light
level at which dark-adapted vertebrate eyes can still see. The only way to
establish the level of illuminance once below -4 log lux is to measure, in
bright daylight conditions, the attenuation of daylight by the water column.
This is done by lowering a sensitive light meter and recording the light level
at each depth. Then, for example at night, the surface level can be measured
and a calculation made either to establish the depth for a particular light
level or to estimate the light level at the study depth. This is the principle
shown by the movable part of Fig. 5.3. The rate of attenuation with depth
can be very fast with turbid water or very slow in crystal clear ocean water
(see example slopes in Fig. 5.3). The rate of attenuation is, however, not
predictable: it commonly changes with depth and must be measured if
behaviour observations are to be related to the real light levels where the
animals are. In order to complete this picture, the level of turbulence-
activated bioluminescence should be monitored. This can be bright enough
in some waters to be filmed at night using the SIT camera (Wardle, 1983;
Glass and Wardle, 1989).

5.2.4 The effects of artificial light

There is nothing more frustrating than finding that there is just not enough
natural light to see when the camera is at the working depth. For behaviour studies there are many reasons why artificial light might be expected to modify observed behaviour. The basis of the argument is not
obvious unless you have dived and observed the underwater view in the
sea. The natural lighting is arranged in a very repeatable and particular way.
The light field is so symmetrical that if a large mirror is suspended vertically
in the sea midwater, it is completely invisible. This is because from any
viewing position the wall of light seen behind the mirror matches exactly
the reflection of the similar wall of light behind the viewer. This wall of
light is bright above and dark below and evenly graded in between, with
all the horizontal levels having a matching intensity. In shallower water
and late and early in the day this spherical symmetry can be tipped slightly
by the sun's position, but in practice not by very much. A sundial would
not work below the surface. Within this very uniform visual framework
objects like fish have adapted to become least visible. Fish even use
vertically mounted mirrors in their scales, making use of the principle
described above.

Silver fish are particularly invisible in their natural light field. However, if
illuminated from the side by a flash or a floodlight, the mirrors in their
flanks, being already lined up, reflect the source back to the camera. Fish use
this property naturally when in a school: individuals are observed to roll on

their side, and this generates a bright flash as the reflection of down-welling light is seen reflected in their flanks, against the darker background below.

Animals found in deeper water have bright red pigments in their skin that are never normally seen as a red colour, but the red pigment becomes black and counter shades their bodies in green-dominated illumination. The addition of even the smallest of artificial light throws out this balance, making objects designed to be least visible shine out against the delicate underwater background. Once the balance is upset, objects appear in quite unexpected ways to both the observer and the other animals present, and any behaviour observed or recorded in these conditions may be expected to be quite unnatural. The more obvious behavioural interference is the presence of the intense light source, having a disorientating effect similar to that observed at a lighthouse on migrating birds and moths. On one occasion, at about 50 m deep at night, the lights were switched on to make visible swarms of shrimps when suddenly in swam large numbers of cod, gulping the otherwise unseen shrimp. The effects are not always so obvious.

5.2.5 Are there camera lights that animals cannot see?

A question often considered is, 'Are there light sources that a particular animal might not see, or can a flash be so short that the animals are not aware of it?'. There are a few studies where it is claimed that the subjects were illuminated, say, with deep red, which it had been shown they could not see. Longer wavelengths of infra-red, such as energy from infra-red diodes used in domestic remote controls, will only penetrate up to 0.5 m even in the clearest seawater, but these sources could be used to show smaller organisms close up and can be used as an invisible source for detecting and counting passing animals, or to trigger a camera system. Silicon diode array and CCD cameras are very sensitive to infra-red wavelengths and are usually fitted with a removable infra-red filter to avoid odd illusions in daylight use. These cameras might have short-range applications with infra-red sources underwater. They are widely used in laboratory-based experiments.

Ultraviolet lights (3500–4000 Angstrom) were claimed to substantially increase the visible range when used by divers at night (Woodbridge and Woodbridge, 1959). Their observations suggest that the particles in the water were not illuminated by the UV, whereas other objects such as coral fluoresced. This combination made the objects with a daylight visibility range of only 2 m visible at 15 m in water. Seawater, where yellow substance is absent, does transmit ultraviolet, but no practical camera observation system has apparently been developed with UV light and these wavelengths are damaging to the eyes. Probably if a camera was selected that was sensitive to UV light then the particles might be very visible in this short-wavelength light.

Work on gated laser systems has shown that, although they are not yet practical, they are very attractive as ideas. The principle is that the light pulse is so short that when it reaches the object being viewed the water between the camera and the object is not illuminated. The camera is arranged (by electronic gating) to open to light only at this instant, and the water between the illuminated object and camera is not seen, giving an unobscured clear image.

An alternative to underwater light is the use of ultrasonic echo sounding (sonar) techniques and, although there are many useful specialist tools of this nature that can add to the camera techniques, there does not yet appear to be a system where the information is updated rapidly enough to visualize faster-moving objects with target strength as delicate as individual fish. The faster the movements the less likely are such systems to be successful. We are still waiting for the large-scale ultrasound camera that will display the dynamic reactions of individual fish moving within the open spaces of a towed trawl. Narrow segments of sea bed up to about 300 m range can be monitored for slow fish movements, using sonar sector scanning systems that have mainly been developed for the needs of oil exploration companies in monitoring bulky vehicle and diver movements.

5.2.6 Can or should cameras be made invisible to fish?

It is well known that a school of fish reacts to the approach of a diver or any intruding object, and the use of a camera to show the natural undisturbed distribution and attitude of the fish beneath or within a volume of the ocean can run into problems. As soon as the camera enters their field of view the fish may react, polarize their movements and change their orientation. The application of mirrors, countershading, the cover of darkness and other tricks may be worthwhile in these conditions. In other situations, perhaps on the sea bed or alongside substantial fishing gears, the camera system may simply become part of the complicated visual background and may not be noticed by the fish. In any circumstances, avoid a camera that makes sounds and cover all indicator lights, either of which might draw attention to its presence.

5.3 A BASIC SYSTEM AND ITS DEVELOPMENT FOR SPECIAL PURPOSES

5.3.1 The best underwater system?

Current options for using TV to record the behaviour of animals underwater can be chosen from domestic, sporting, military, scientific and commercial developments: domestic systems are getting more and more useful as recording devices. There are now many sport-diving video systems that work well in shallow scuba diving situations. For work in deeper water, the basic decisions involve making the most use of natural light where possible while realizing that colour becomes less and less important with increasing depth. Beyond about 30 m depth the only camera tubes that work with natural light are the SIT and ISIT types.

The most economical arrangement is a self-contained camera and recording system where the video tape is recording while the camera is immersed, and the recordings are only seen when the system is hauled out, opened and replayed. Alternatively, the recorder and monitor may be in a boat or on the shore and the camera on the sea bed linked by a cable. The camera could be linked to the surface raft or buoy by cable, and a radio transmitter unit used to send the TV signal as far as 6 km away to a shore-based laboratory. At sea, with big ships, systems can be lowered while at anchor or while drifting or can be adapted to be towed over the stern. Cameras may be carried by specially designed vehicles with many levels of sophistication of winch systems and remote controls, or human divers may be involved. We will consider some examples where these different systems have been applied.

5.3.2 Marine video systems

The principal components of an underwater observation system using TV are outlined below. It should be realized that hand-held camcorders which are becoming common as domestic and amateur systems can contain some form of these components all in one compact unit.

The camera

The camera should be compatible with the expected light level range. The best camera type for marine conditions has repeatedly been proved to be a camera fitted with the SIT type of tube. The camera is usually more useful with an automatic light level response so that it operates safely and correctly in bright and dark conditions without attention. Some experiments might need the camera preset, or the ability to adjust the sensitivity manually. As with all cameras the lens aperture limits light gathering and the focal length determines the field of view. In many underwater situations short-range viewing is the more normal condition due to poor water clarity, and the widest angle lens (shortest focal length lens) is the most useful. Zoom or long focal length lenses are rarely needed but might be useful in very clear water. The quality of the picture in general is limited by the tube quality, and high-quality lenses as used in cine cameras are unlikely to improve on these limits. The camera needs to be housed within a pressure vessel rated to the working depth, and there are various window designs and lens attachments that correct for distortion due to the air–water interface. A copper ring around the window can reduce fouling of the glass by organisms.

Surface control units

The underwater TV camera is often separate from its power supply, display monitor, video recorder, video clock and camera controls, which are all kept in a dry place at the surface. Surface systems may be integrated in a damp-proof portable case including all the essential units, or may be preferred as more easily serviced and replaced separate items linked by cables. More complex units may have additional features such as video overlay units (the facility to merge text, graphics or other video sources with the prime source), real-time digital clocks, data monitors and auxiliary controls, all contained within a small rack suitable for easy transportation.

TV cables

A multicore waterproof cable is needed to link the camera to the surface units. The cable can have extra conductors to control lights, and a pan and tilt system is needed to adjust the pointing of the camera by remote control. The cable is not usually strong enough to support any weight on the lower end, so may need to be used in conjunction with a rope or stranded wire if a heavy camera frame or negatively buoyant towed body is to be deployed. Depending on the diameter and weight or buoyancy of the cable, it may not be possible to hold it by hand if more than about 40 m are deployed unsupported. Heavier cables are best attached to a lifting wire worked from a winch. Practical attachment is either by tying on or using some non-slip arrangement such as miniature 'Eureka' wire grips on the lifting wire and karabiner clips on the cable. This type of system is simple to attach and release, but the deployment of greater lengths may take a long time and is physically demanding. It is better to use an umbilical cable with a strain-bearing core such as Kevlar, a non-stretch material lighter than steel, or an armoured high-tensile sheath. These cables must be deployed from a winch because of the increased cable weight and load, but heaving and shooting the cable becomes a simple matter. An electrical slip-ring assembly is required in the centre of the winch to enable the cable conductors on the moving drum to be in connection with the surface recording and control units. If the camera system is deployed with a remote vehicle which has the ability to hover, such as the Robertson SPRINT ROV, it may be possible to use a neutrally buoyant cable and pay out manually, as there will be little or no load on the surface end of the cable. Cables are always vulnerable to physical damage, and those handling them need special training. They can be a major cost in running underwater systems.

5.4 RADIO LINKS FOR REMOTE CAMERAS

In locations where a cable connection to the camera is impractical, due to excessive distances between camera and shore station, or intervening underwater obstacles such as deep gullies, it is possible to use a video transmission system ('Vision Link' by Optical and Textile Ltd.). Here, the camera is isolated with a suitable power supply, and the video signal transmitted via radio link to the receiver. Distances of up to 6 km may be covered, which would be difficult to achieve by a conventional cable link. In practice, the camera power supply – for instance a car battery – is mounted in a small raft or buoy where it is accessible for servicing. The radio system can be used in both directions, enabling the remote control of pan and tilt units, so that virtually any system workable with cable links may be replaced by a radio-based system. Such a system is easier to deploy and less prone to damage than a cable system.

5.5 VIDEO RECORDING

Video recording can be real-time or time-lapse. The portability of the system might be more important than the quality of the recording. Quality is usually reflected in the preservation of a high-frequency response in the recorded signal that might be expressed in how rapidly the scanning beam on the TV screen can change from black to white and back again: the more sluggish this gets, the lower is the resolution. The cameras used are relatively low quality already in this sense, and most of the available recording systems will preserve the camera's ability. However, editing and copying tests this ability to preserve the frequency response and the video tape needed for this purpose is better made on the bigger formats with the higher-frequency response. U-Matic low-band systems, for example, adequately preserve the picture quality but the normal tapes only record 1 hour. VHS systems give longer recording times and there are superior SVHS systems. 8 mm systems are extremely compact and often incorporated in the portable scuba diving cameras. Films made on the more compact systems can, of course, be copied to better systems that then help to preserve the quality in editing. It is a sensible precaution when using generators and other portable power supplies at sea or in remote recording sites, to filter the supply through a stabilizer to prevent damage to the recorder or interference on the tape.

5.5.1 Some examples of successful observation in the sea

Figure 5.4 shows an experiment set up in the sea close to a shore laboratory to observe the reactions of wild fish to a length of small line (Johnstone and Hawkins, 1981), where action away from the main area of interest may be

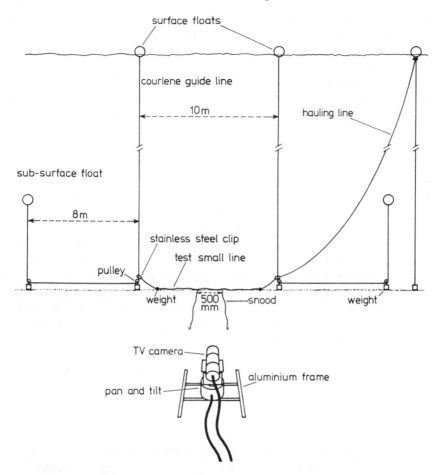

Figure 5.4 An experimental arrangement of camera and pan and tilt for viewing a test length of fishing small line (from Johnstone and Hawkins, 1981).

viewed by moving the camera. Cameras suspended just above the sea bed or on a framework have been used by Schwenke (1965), Farrow *et al.* (1979) and Goeden (1980), and fixed cameras have been used for long-term observation by Nishimura (1966), Chapman and Rice (1971) and Eleftheriou and Basford (1983). Studies of *Nephrops* by Chapman and Howard (1979) showed the importance of using light filtered to be outwith the sensitivity of the lobsters.

5.5.2 Towed bodies and sledges

Conventional methods of surveying larger areas of midwater or sea bed involve towing a frame containing instruments, sampling devices or cameras, and many specialist towed-body systems have been developed; some of these have been developed further in order to make ecological and behavioural observations.

One form of towed-body sledge is fitted with a TV camera as a monitor that allows the more effective manual operation of high-quality photographic flash cameras, and was described by Chapman (1985). The main features of this sledge (Fig. 5.5) were the TV and photographic cameras arranged to observe the distribution and behaviour of *Nephrops* with the least disturbance. The TV camera view was illuminated by red-filtered artificial light known to be invisible to this species. The particular camera arrangement allowed photographs of the organisms of interest to be taken just as they passed the centre of the camera field. The overlap of the TV and photographic camera fields was such that the bottom edge of the television screen corresponded to the centre of the Hasselblad film camera field. This arrangement gave the operator about 3 s warning (Fig. 5.5, distance *D*) as he watched the subject arriving at the point where a centred photograph was taken by pressing a switch. Measurement of the dimensions of the subjects in the camera field of view was achieved by reference to the width of the tracks of the runners of the sledge left in the soft sea bed. An alternative system has more recently been used where a pair of laser projectors are mounted on the sledge, such that two bright spots a known distance apart appear on the sea bed in the recorded views. There are many variations on this theme, and other sledges for sea-bed use are described by Machan and Fedra (1975), Holme and Barratt (1977) and Bascom (1976).

Scuba diver's towed vehicle

Main and Sangster (1983) describe a towed submersible operated by divers, which overcomes most of the problems experienced in using TV cameras to film the behaviour of fish in towed fishing gears. In this fast-moving situation the shelter of the scuba divers' vehicle (Fig. 5.6a) makes it possible to support the towed TV cable and maintain a precise position alongside or in the trawl net, while working the camera effectively. This optimizes the use of the limited diving time available to each diver in a day. The vehicle is towed by the fishing vessel and moved relative to the gear by hauling or letting out the towing cable (Fig. 5.6b). This system is worked with both TV and voice communication between the towing vessel and the submersible, enabling real-time pictures to be taken of specific points of interest by immediate discussion between surface and diving scientists.

Remote-controlled towed TV camera vehicle

Observations at fishing depths of 30 m or more cannot be undertaken by manned vehicles because decompression times for human divers would be

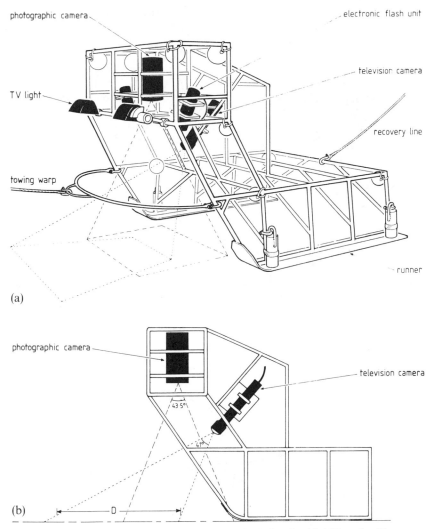

Figure 5.5 (a) A sledge for towing on the sea bed, fitted with TV and still cameras. (b) The arrangement of cameras on the towed sledge to enable distance and speed measurements to be made, showing camera fields allowing for sinking of runners into the sand. Distance D ($= 1.03$ m), corresponding to the height of the TV picture, was used to estimate towing speed and distance. (From Chapman, 1985.)

impractical. Based on work by Magnus (1852), Smith and Stubbs (1970), Silva and Ferro (1974) and Edwards (1981), a remote-controlled vehicle (RCTV) was developed specifically for the observation of fishing gear and fish behaviour, as described by Priestley *et al.* (1985) (Fig. 5.7). This remote-controlled system is now widely used in fishery laboratories around the world for the observation of the behaviour of fish reacting within the vicinity

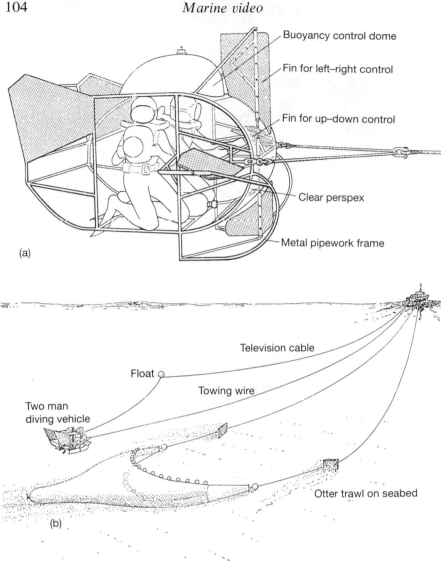

Buoyancy control dome

Fin for left–right control

Fin for up–down control

Clear perspex

Metal pipework frame

(a)

Television cable

Float

Towing wire

Two man
diving vehicle

Otter trawl on seabed

(b)

Figure 5.6 (a) A towed submersible for operation by divers. (b) Diagram showing how it is towed. Both as used at the Marine Laboratory, Aberdeen.

of towed commercial trawl nets. The results from this research in the form of video tapes and published observations and interpretations are stimulating the development of more precise and selective fishing gears, as it has been found that the observed behaviour patterns are often species-specific. The remote vehicle systems take the SIT camera to its limit of sensitivity at about 100 m depth in clear water. There are tricks that can be used to extend behaviour observation to deeper or darker conditions.

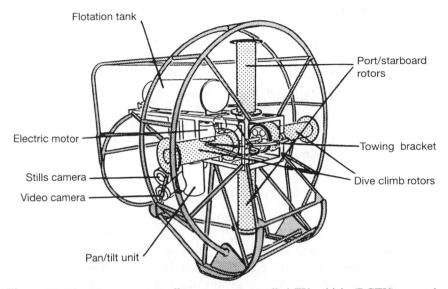

Flotation tank

Port/starboard rotors

Electric motor

Towing bracket

Stills camera

Dive climb rotors

Video camera

Pan/tilt unit

Figure 5.7 The Magnus rotor effect remote controlled TV vehicle (RCTV) as used at the Marine Laboratory Aberdeen for filming the behaviour of fish in fishing gears.

5.6 WORKING IN THE DARK

Experiments have been made using the RCTV to observe fish at light levels close to and below their visual threshold (Glass and Wardle, 1989). As this is beyond the viewing capacity of video cameras, as explained earlier, other means of observing the fish had to be found. At night, floodlights were needed in order to manoeuvre the RCTV alongside the ground-gear of the net. Once in position the lights were extinguished, and flash photographs taken every 20 s or so by manual command from the surface, using a 35 mm camera mounted on the RCTV. The position of the RCTV was checked at intervals by switching on the floodlights again. This lights-on/lights-off 'flying-blind' routine was repeated many times. The behaviour of fish in the viewing field of the stills camera was only known days later, after developing the films. An alternative method involves mounting a sector-scanning echo-sounder system (Simrad FS3300) on the remote vehicle. This then allows the position of the RCTV relative to the net and the sea bed to be constantly monitored and maintained in complete darkness, when nothing is visible with the TV camera. The scientist controlling the position of the vehicle views and responds to a radar-like display. Extended 'lights-out' periods may be carried out safely, allowing well-positioned flash-camera recording of fish reactions to the gear components in undisturbed low light levels.

An experimental development of this approach makes use of the TV camera and flash gun to present immediate flash images to the observer. The TV camera, with sensitivity previously adjusted to operate when the flash-gun is fired, transfers the image as a TV frame to the monitor on the ship above, where it can be photographed with a Polaroid camera or video taped or grabbed by a computer imaging system. The image can be examined immediately, allowing the experiment to develop more rapidly. The development of this technique might be towards a high-resolution TV system, preserving greater detail in the camera image, and colour would help with species identification.

5.7 WORK WITH TV AT GREAT DEPTHS

Behaviour experiments have now reached very deep water. Using a specially developed free-fall vehicle (Fig. 5.8), Priede and Smith (1986) and Priede *et al.* (1990) were able to observe grenadier fish on the sea bed at 5400 m deep. They were able to watch the fish approach and swallow bait containing an acoustic transmitter tag. After ingesting the tag the animals' activity and movements in the vicinity could be tracked. The camera and the video recorder were together in an ATEX sphere and the video tape was analysed once recovered from the depths.

5.8 SELF-PROPELLED CAMERA-CARRYING VEHICLES

A range of unmanned remotely operated vehicles (ROVs) fitted with TV cameras are available in the commercial market, largely due to the great demand for their particular abilities in respect of economic underwater equipment inspection at oil exploration sites. These vehicles are compact self-propelled vehicles that can be three-dimensionally positioned while hovering at the observation position. One owned by the Marine Labora- tory at Aberdeen has been used in herring larvae distribution surveys, investigation of salmon migration in relation to hydroelectric dams, studies of the Norway lobster in their burrows on the sea bed, and have played a key role in investigations of the role of sea-bed disturbance on the foraging success of juvenile fish (Hall, S.J. unpublished observations). In this forag- ing experiment, it was a requirement to obtain specimens of the fish after observing their feeding routines. A net system attached to an ROV was developed to capture specimens observed feeding in the test area. A framework was built from lengths of 50 mm diameter aluminium scaffold- ing tube held in place by quick-release right-angle clamps to form a cube 1.40 m on each side. The support skids of the ROV were bolted to the top of the frame in such a position that its camera viewed the base of the cube resting on the sea bed. By lowering a remotely controlled motor- ized rotating disc against the sea bed the sediment is disturbed. After

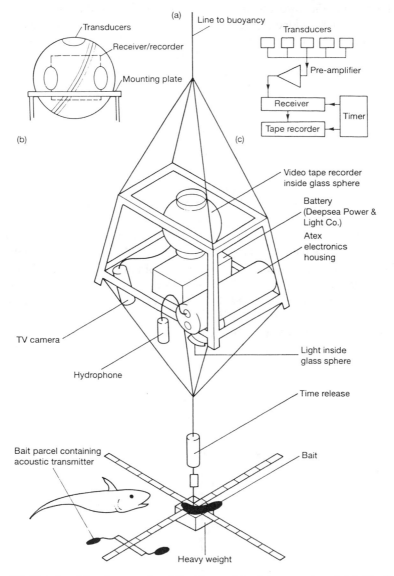

Figure 5.8 Details of the deep-sea free-fall video and tracking experiment. (a) The free vehicle video acoustic tracking experiment (FVV/ATEX). Acoustic transmitters wrapped in balls of bait can be fitted either on the V bridles below the vehicle or on the cross, as indicated. Each arm of the cross is 1 m long. The camera is in the lower of two placements used, 1.5 m above the cross. (b) The ATEX sphere, a conventional 25.4 cm diameter glass deep-sea instrument housing, with all acoustic sensing elements bonded to the glass and all the equipment fitted within the sphere. (c) Schematic of the ATEX receiving and logging system. Dry-cell batteries were also fitted within the sphere (for more details see Priede and Smith, 1986, and Priede *et al.*, 1990).

withdrawing the disc the approach and feeding of the fish is observed. Netting panels folded up on the sides of the frame were released by command from the experimenter watching the TV screen. These quickly fall to form a curtain around the four sides of the frame, and another panel is drawn across the base, trapping any fish which were inside the frame at the time.

5.9 USE OF CAMERAS TO MEASURE LENGTH

One of the properties of a TV camera with a fixed lens system is that the length of an object can be measured on the screen as a pair of *xy* coordinates, using apparatus like the Hampton Video VP112 digitizer, but the range of the object must be accurately known. By using a high-precision echo-sounder in water, the range of objects in the centre of the camera view can be measured easily with a 1 cm resolution at up to 8 m. The advantage of very high-frequency sonar (8 mHz in the Microranger by Seametrics, Aberdeen) is that the transducer is very small and easily attached alongside the camera, and the sampling beam is very narrow and precise. Bluefin tuna, swimming horizontally above a TV camera arranged to look vertically up at them (Fig. 5.9), were measured in this way at a remote field site, while their swimming movements were being recorded. These fish were in a situation where they could not have been measured in any other way (Wardle *et al.*, 1989).

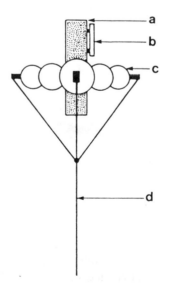

Figure 5.9 A vertically mounted TV camera with rangefinder to film and measure blue fin tuna swimming above. a = TV camera; b = rangefinder; c = 200 mm floats on a stiff collar; d = line to heavy weight on sea bed (from Wardle *et al.*, 1989).

The use of stereo-pair cameras for measurement has been developed both commercially and by scientists for some specialist purposes. The published results are rare, and perhaps indicate that this route to accurate measurement, although theoretically attractive, can be difficult to apply and is less practical than other more straightforward measurement systems.

5.10 CONCLUSIONS

TV and video equipment are valuable tools in many areas of subsea research, enabling real-time viewing, sampling and measurement of events at remote sites, and providing the facility for long-term storage of recorded material. However, there is a wide choice of very expensive equipment available, and its use at sea can be particularly demanding in resources and manpower. Subsequent maintenance and operation needs the support of specialists with technical skills, and great care is needed in selecting the most appropriate and practical system to use. Reference to professional and experienced users is essential.

REFERENCES

Barnes, H. (1963). Underwater television. *Oceanography and Marine Biology, An Annual Review*, **1**, 115–128.

Bascom, W. (1976). An underwater television system. Southern California Coastal Water Research Project. *Annual Report, 1976*, 171–174.

Chapman, C.J. and Rice, A.L. (1971). Some direct observations on the ecology and behaviour of the Norway lobster *Nephrops norvegicus*. *Marine Biology*, **10**, 321–329.

Chapman, C.J. and Howard, F.G. (1979). Field observations on the emergence rhythm of the Norway Lobster *Nephrops norvegicus*, using different methods. *Marine Biology*, **51**, 157–165.

Chapman, C.J. (1985). Observing Norway lobsters, *Nephrops norvegicus* (L.) by towed sledge fitted with photographic and television cameras. In *Underwater Photography and Television for Scientists* (eds J.D. George, G.I. Lythgoe and J.N. Lythgoe). Clarendon Press, Oxford, pp. 100–108.

Cui, G., Wardle, C.S., Glass, C.W. *et al.* (1991). Light level thresholds for visual reaction of mackerel, *Scomber scombrus* L., to coloured monofilament nylon gillnet material. *Fisheries Research*, **10**, 255–263.

Edwards, W.G. (1981). Patent Specification Number 1596275 (54) *Improvements in or Relating to Underwater Vehicles*. 1–8 and 17 figures.

Eleftheriou, A. and Basford, D.J. (1983). The general behaviour and feeding of *Cerianthus lloydi Gosse* (Anthozoa, Coelenterata). *Cahiers de Biologie Marine*, **24**, 147–158.

Farrow, G., Scoffin, T., Brown, B. and Cucci, M. (1979). An underwater television survey of facies variations on the inner Scottish shelf between Colinsay, Islay and Jura. *Scottish Geological Journal*, **15**, 13–29.

George, J.D., Lythgoe, G.I. and Lythgoe, J.N. (1985). *Underwater Photography and Television for Scientists*, Clarendon Press, Oxford.

Glass, C.W. and Wardle, C.S. (1989). Comparison of the reactions of fish to a trawl gear, at high and low light intensities. *Fisheries Research*, **7**, 249–266.

Goeden, G.B. (1980). Reef survey finds commercial fish habitats. *Australian Fisheries*, **39**(6), 8–9.

Harris, R.J. (1980). Improving the design of underwater TV cameras. *International Underwater Systems*, **2**, 7–11.

Holme, N.A. and Barratt, R.L. (1977). A sledge with television and photographic cameras for quantitative investigations of the epifauna on the continental shelf. *Journal of the Marine Biological Association of the United Kingdom*, **57**, 391–403.

Holme, N.A. and McIntyre, A.D. (1985). *Methods for the Study of Marine Benthos*. IBP Handbook 16, 2nd edn, Blackwell Scientific Publications, Oxford.

Johnstone, A.D.F. and Hawkins, A.D. (1981). *A Method for Testing the Effectiveness of Different Fishing Baits in the Sea*, Scottish Fisheries Information Pamphlet No. 3.

Machan, R. and Fedra, K. (1975). A new towed underwater camera system for wide-ranging benthic surveys. *Marine Biology*, **33**, 75–84.

Mackay, D. (1991). *Underwater Television Sensors* – Selection Criteria, Osprey Electronics Ltd., Campus 1, Balgownie Rd., Bridge of Don, Aberdeen.

Magnus, H.G. (1852). Uber die abweichung der Geschosse und eine auffallende erscheinung bei roirenden Korpern. Abhandlungen der Akademie der Wissenschaften in Goettingen zu Berlin.

Main, J. and Sangster, G.I. (1978). *The value of Direct Observation Techniques by Divers in Fishing Gear Research*, Scottish Fisheries Research Report No. 12.

Main, J. and Sangster, G.I. (1983). *TUV II – a Towed Wet Submersible for Use in Fishing Gear Research*, Scottish Fisheries Research Report No. 29.

Nishimura, M. (1966). A study of the application of underwater television (Report No. 2). *Technical Report of Fishing Boat*, **20**(4).

Priede, I.G. and Smith, K.L. Jr. (1986). Behaviour of the abyssal grenadier, *Coryphaenoides yaquinae*, monitored using ingestible acoustic transmitters in the Pacific Ocean. *Journal of Fish Biology*, **29**, Suppl A, 199–206.

Priede, I.G., Smith, K.L. and Armstrong, I.D. (1990). Foraging behaviour of abyssal grenadier fish: inferences from acoustic tagging and tracking in the North Pacific Ocean. *Deep Sea Research*, **37**, 81–101.

Priestley, R., Wardle, C.S. and Hall, C.D. (1985). *The Marine Laboratory Remote Controlled Fishing Gear Observation Vehicle*, ICES CM1985/B:10 Fish Capture Committee, pp. 1–7 and 9 figures.

Schwenke, H. (1965). Uber die anwendung das unterwasserfernsehens in der Meeresbotanik. *Kieler Meeresforschungen*, **21**, 101–106.

Silva, S.T.R. de and Ferro, R.S.T. (1974). *A Study of the Use of Power Rotors to Improve the Manoeuvrability of Pelagic Trawl Gear*, ICES CM1974/B:10 Gear and Behaviour Committee, pp. 1–3 and 5 figures.

Smith, K.E. and Stubbs, H.E. (1970). Rotor device for controlling depth of towed fishing trawls, in *FAO Technical Conference on Fish Finding, Purse Seining and Aimed Trawling*, FAO, Rome, pp. 1–13.

Tyler, J.E. and Smith, R.C. (1967). Spectroradiometric characteristics of natural light underwater. *Journal of the Optical Society of America*, **57**, 595–601.

Wardle, C.S. (1983). Fish reactions to towed fishing gears, in *Experimental Biology at Sea* (eds A. MacDonald and I.G. Priede), Academic Press, London, pp. 167–195.

Wardle, C.S. (1985). Investigating the behaviour of fish during capture. In *Developments in Fisheries Research in Scotland* (ed. R.S. Bailey and B.B. Parrish), Fishing News Books Ltd, Farnham, England.

Wardle, C.S. (1993). Fish behaviour and fishing gear, in *Behaviour of Teleost Fishes*, 2nd edn (ed. T.J. Pitcher), Chapman & Hall, London, pp. 609–643.

Wardle, C.S., Videler, J.J. Arimoto, J.M. *et al.* (1989). The muscle twitch and the maximum swimming speed of giant bluefin tuna, *Thunnus thynnus* L. *Journal of Fish Biology*, **35**, 129–137.

Woodbridge, R.G. and Woodbrige, R.C. (1959). Application of ultra-violet lights to underwater research. *Nature (London)*, **184**, 259.

6

Wild birds

K.W. Smith

6.1 INTRODUCTION

Although video-recording equipment is now familiar in the majority of homes and security cameras and monitors adorn many shops and garages, it is surprising how little the technology has been applied to work on wild birds. The first reported use of video equipment in studies of avian ecology was by Haftorn (1972). He used closed-circuit video cameras to study the breeding behaviour of a number of bird species in both nest boxes and open nests. Considering the novelty of the equipment at the time, this was a remarkable and far-sighted innovation. Since then, there have been rather few reported uses of video techniques *per se*, although there are probably many instances where such equipment has been used but not reported.

Since Haftorn's work, the range of available equipment and its capabilities have increased significantly and the real costs have fallen, making video the best solution in many applications. This chapter will compare the performance of video systems with conventional photographic techniques, review previous work with wild birds and present examples of the use of video from conservation studies of wild birds.

Conventional photographic systems have been used for a variety of purposes but usually at nests or other places where the birds occur predictably. The simplest approach has been to use time-lapse cameras to monitor nests or breeding colonies over prolonged periods to collect data on attendance patterns, feeding rates, diet and causes of failures (Mudge *et al.*, 1987, Tommeraas, 1989).

In more sophisticated applications cameras triggered to take pictures of particular events have been used. For instance, cameras that take a single picture each time an adult returns with food have been used extensively in

Video Techniques in Animal Ecology and Behaviour. Edited by Stephen D. Wratten. Published in 1993 by Chapman & Hall, London. ISBN 0 412 46640 6

studies of species using nest boxes (see for example Perrins, 1979). Triggering devices include sensitive perches, trip wires and light beams interrupted by the passage of the bird. Similar systems have been used to identify predators at real and false nests (Hussell, 1974; Picman, 1987; Savidge and Seibert, 1988; Major, 1991). Although video systems could potentially be used in all these applications, it would probably not be sensible to do so. Other factors such as cost, image quality and power requirements need to be taken into consideration in selecting the best approach. The advantages and disadvantages of video compared with conventional photographic systems are set out in Table 6.1.

Haftorn (1972) used video cameras and other data-logging techniques to study the breeding biology of 12 species of birds, both open nesters and hole nesters. Cameras were hard wired back to a study hut where images from all the nests could be viewed on a monitor and, where necessary, stored on a video recorder. Power was provided by a portable generator at the study hut. The system was used to study incubation behaviour in relation to nest temperature by experimental manipulation of the temperature within the nest boxes using electric heaters (Haftorn, 1984). Although most of these studies involved birds using nest boxes, Haftorn also used video techniques to study nesting goldcrests, *Regulus regulus*, and described the details of the egg-laying and fledging sequences (Haftorn, 1978a,b,c). In using video he was able to exploit the ability to transmit images back to a base station where they could be viewed in real time with no disturbance to the birds.

Table 6.1 The advantages and disadvantages of video systems compared with conventional photography

---

Advantages
Video cameras typically operate to lower light levels than photographic systems and can more easily be used with infra-red illumination

Video tapes hold far more images than conventional cine film, i.e. approximately 250 000 frames on a standard video tape compared with around 4500 on an 8 mm super-8 film

Video tapes can give instant playback

Being an electronic image video can be transmitted away from the place of collection, either for a limited distance by cable or throughout the universe by radio transmission

Disadvantages
Video has lower image quality than photographic systems

Video systems require a power source, which can cause problems if prolonged operation in remote locations is required

Video is usually more expensive than photography, although the initial capital cost has to be offset by the higher cost of films and processing

---

Video cameras have been used routinely to monitor the events in aviaries during captive breeding programmes, but it was not until 1983 that systems were described that could be used at 'hack' sites away from base, or at the nests of wild birds used to adopt captive-bred young (Wisniewski, 1983). In his work on the reintroduction of bald eagles *Haliaeetus leucocephalus* and peregrine falcons *Falco peregrinus*, Wisniewski used black and white and colour cameras hard wired for power and video back to a convenient point which could be approached without disturbing the birds. Wardening staff could simply plug into the video line, power up the camera and view the current status at the nest. This was far more efficient than the conventional system of an observer having to enter a hide during the hours of darkness to observe the birds the next day and remain there for the whole day. Cable runs of up to 800 m were used and power provided by lead–acid batteries. Although the cameras were initially used by staff to monitor events at the sites, the additional potential for publicity was quickly recognized and exploited. Viewing sites were established where the public could watch video images of events at the nests, and video tapes were provided to the local media to promote interest in the project.

The public viewing potential of video systems has now been widely exploited. On many nature reserves it is now possible to watch video monitors of events not easily visible from the hides – usually bird nests or closer views of bird flocks and roosts. In the UK the RSPB has been using such systems for many years, mainly in cliff-nesting seabird colonies, where it is often difficult or dangerous for the public to see the birds.

One of the earliest uses of video for such purposes in the RSPB was 'Operation Woodpecker' in 1987, in which the public were shown live video of the events inside the nesting cavity of a pair of great spotted woodpeckers, *Dendrocopus major*, while simultaneously being able to watch the nest site from a nearby hide (Davies and Campey, 1987). This approach was sufficiently novel to attract considerable media attention and, over the 3 weeks of operation, to attract over 15 000 visitors. The use of real-time video is now widespread on reserves, although the concept of visiting a nature reserve to watch a television monitor does not suit everyone.

Wisniewski described the use of a radio link rather than coaxial cable to transmit video images back to a base station. Although cables can be used easily up to a distance of 1 km or so, beyond this there are problems with loss of image quality and logistical difficulties in routing cables around, under or over obstructions. Technically there are no particular difficulties in setting up a radio link, and many such systems are available from security companies. In practice, however, in most countries, restrictions on available radio frequencies severely limit the possibilities. Video signals need a wide bandwidth and, although the UHF band is adequate, in the UK no such frequencies are made available for this type of use. Currently the only

frequencies allocated for video transmission in the UK which could potentially be used by ecological researchers are in the microwave band. The use
of these frequencies would necessitate expensive equipment and limit the
link to strict line of sight, with all the associated alignment problems this
would bring. Various optical links, such as lasers or focused beams, are
available which avoid the problems of licensing, but these are almost as
expensive as microwave systems, have limited range, do not work in
conditions of low visibility such as fog or heavy rain, and are extremely
sensitive to alignment. In spite of these problems they are now widely used
for short-range links between buildings in security installations and computer networks.

The advent of miniature CCD cameras has created a number of novel
applications. Dyer and Hill (1991) used such a camera instead of a fibreoptic
endoscope for inspecting burrow-nesting seabirds. They used a miniature
black and white camera head with umbilical cord back to the controlling
electronics (Panasonic WV-CD50). Their camera had some sensitivity in the
infra-red, and so an array of infra-red light-emitting diodes was used for
illumination. The whole system, including a TV monitor, was powered by
batteries. The advantages compared with a standard endoscope were lower
cost and power consumption and, because of the lower levels of illumination
that were required, potentially less disturbance to the occupants of the
burrows.

Miniature camera heads are now finding application in studies of hole-
nesting birds in natural cavities (Koenig, pers. comm.). The size of conventional cameras makes their installation inside natural nesting cavities
extremely difficult and potentially time-consuming, whereas miniature
cameras only require a small hole in the cavity.

Although camcorders have been on the market for some years, the
literature contains few specific references to their use in bird studies. They
could clearly be used in any studies where cine cameras had been used
previously, such as behavioural and nest studies. They have the advantages
of longer run-time per cassette and low film cost. The disadvantages are high
initial cost (although this is falling), significant power consumption, unpro-
ven robustness in the field and sensitivity to adverse environmental conditions. Some camcorders have the facility to take time-lapse sequences. This
is achieved by taking successive short bursts of video, rather than the single
frames of a conventional time-lapse video recorder. They are thus not able
to record such long time-lapse sequences on a single cassette. In any case,
their power consumption would become a significant problem in any remote
long-sequence time-lapse work.

To illustrate some of the questions that can be addressed by using video
techniques, and to highlight some of the problems that have to be overcome,
two pieces of work on threatened species in the UK will be described: the
red kite *Milvus milvus* and the cirl bunting *Emberiza cirlus*.

6.2 BREEDING FAILURES OF RED KITES IN WALES

The red kite is a rare and threatened breeding bird, with its world distribution largely restricted to Europe (Batten *et al.*, 1990). In Britain its range is restricted to a small area of Wales, where the population has increased from a remnant of as few as five pairs early this century to the current level of around 100 territorial pairs (Davis and Newton, 1981; Davis, pers. comm.). Although there has been a consistent increase in the population since the 1940s, the rate of increase has been low (Davis and Newton, 1981) and has only been sustained with the help of expensive conservation and protection measures (Lovegrove, 1990). One of the factors limiting the growth of the population has been the consistent low breeding success of the Welsh birds which, until recently, has been around half of that of red kites in the rest of their range (Davis and Newton, 1981). Many reasons have been suggested for this, but the most feasible is poor food supply, possibly affecting incubation patterns or chick survival (Davis and Newton, 1981).

Breeding red kites in Wales are particularly sensitive to disturbance, and so visits to nests have always been limited and few data were available on the exact circumstances of breeding failures. If steps were to be taken to improve their productivity, it was clear that more information was required and potential strategies tested. We were therefore asked to devise methods of investigating this problem without causing disturbance to the birds.

After some initial work involving false eggs to monitor attendance patterns during incubation (Smith, 1987, 1988), it was clear that the way forward was the use of video cameras to film the events at the nest throughout the nesting cycle. It was obviously unacceptable to install video cameras at active nests but, fortunately, red kites are very faithful to their nesting woods from year to year, and either reuse their old nest or build a new one nearby (Walters-Davies and Davis, 1973). Our strategy was therefore to install a large number of cameras at potential nest sites in January and February, before the birds were on territory, and to rely on some of the nests actually being used (Lovegrove *et al.*, 1990). Overall, around 25% of cameras were installed at nests which were eventually used by the birds.

Cameras were 12 v black and white CCD (Sony AVC-D5CE or Crofton 801) and at installation cables were buried back to a base station where a time-lapse video recorder (Mitsubishi HS-480E) could be installed. To avoid any possibility of disturbance, the base station was usually at least 200 m from the nest site. The recorder was installed in either a weatherproof housing or, in many instances, in a farm building from where mains power was often available. The maximum cable run was 800 m. Cables were buried or run along fence lines, both to avoid damage by grazing animals and to avoid drawing attention to the nest site.

It was essential that the system could run without attention for long periods, and that tapes could be replaced without disturbing the nesting

birds. The time-lapse recorders could operate for up to 20 days on a single
3-hour cassette, but in practice 10 days was normal and, during interesting
phases of the nesting cycle the period was reduced to 3 days, giving around
two frames per second.

In a few cases it was not possible to find a source of mains electricity
within a reasonable distance of the base station, and so power for the
recorder was supplied from lead–acid batteries via a 12–240 v inverter (RS
Components). The camera was supplied directly from the battery. Such
installations required regular attendance and, with a 110 Ah a battery,
would operate for around 3 days between charges. To avoid too much
carrying of heavy batteries, such installations were best set up close to points
of vehicular access.

The study still continues, but by 1991 data had been collected from 16
nesting attempts (Smith *et al.*, 1991). The outcomes of these are shown in
Table 6.2. Of the 42 eggs laid only 32 (76%) hatched, but the majority of
the losses were the result of the total failure during incubation of two nests.
At one all three eggs were infertile and the other was deserted, one of the
adults having been found below the nest, poisoned. If this incident is
excluded, then 82% of the eggs hatched. Many losses in fact occur after
hatching, with only 59% of young that hatch surviving to fledging. They die
either soon after hatching or within 10–20 days, often after considerable
aggression from their siblings. From the video records data on feeding rates

Table 6.2 The outcomes of 16 red kite nests in Wales between 1988 and 1991,
monitored using video cameras

Nest	No. of eggs laid	No. of young hatched	No. of young fledged
1	2	1	1
2	3	0	adult poisoned
3	3	3	1
4	2	2	1
5	2	2	1
6	3	3	0
7	3	1	1
8	3	3	1
9	2	2	2
10	3	3	3
11	3	3	3
12	2	2	1
13	3	0	0
14	3	3	2
15	2	2	1
16	3	2	1
Total	42	32	19

were extracted for each sibling. A typical example is shown in Fig. 6.1, where the total number of feeding bouts per day received by each of two young in a nest are plotted against date. There is considerable daily variation, but in general the younger of the chicks is fed less often than its older sibling, until it eventually dies. In common with many other birds of prey, red kite young hatch asynchronously so, if there is any competition for food, the younger chick is usually at a disadvantage (Newton, 1979). Also shown in Fig. 6.1 is the number of attacks each chick receives from its sibling. In this example the younger chick is attacked so often that it eventually succumbs.

This evidence supports the view that food shortage is a problem for breeding red kites in Wales. To test this, and to investigate ways of improving productivity, a supplementary feeding experiment is now under way. It has been possible at some nests to halt sibling aggression by providing extra food, and thus to allow more than one chick to survive. There are, however, still difficulties in getting the supplementary food to the

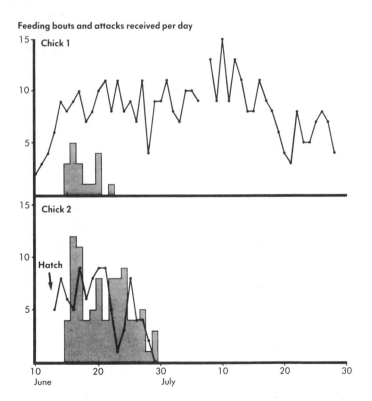

Figure 6.1 The number of feeding bouts per day received by each of two sibling red kite chicks (continuous line) and the number of attacks each received from the other chick (shaded). Chick 2 hatched 3 days after chick 1, and eventually died on July 29th.

birds without causing other problems, such as disturbance or disruption by corvids attracted to the food. Work continues to find ways of overcoming these difficulties.

6.3 FEEDING ECOLOGY OF NESTING CIRL BUNTINGS

The cirl bunting is a small seed-eating passerine with a declining population of 100–200 pairs in the UK, restricted to a small area of South Devon (Evans, 1992). It has been the subject of intensive field studies by the RSPB (Evans, 1991) and a series of actions to bring about a recovery of the population. Although much of this has centred on the provision of winter feeding areas, there has also been the need to identify the key requirements in the breeding season, and to this end investigations of the foods brought to chicks and adult foraging areas have been carried out. Chick diet had already been studied by faecal analysis and direct observation, and was found to be a mixture of seeds and invertebrates (Evans, 1991). As part of a detailed study of foraging habitat requirements, it was decided to use video cameras at nests coupled with direct observations of the foraging adults to determine what foods were being brought from where. The advantage of this system was that observers could concentrate on following the movements of the adult birds, while the video camera recorded events at the nest.

Cameras were installed with a good view of the young in the nest, and hard wired to a video recorder concealed approximately 100 m from the nest in a weatherproof housing. Power was provided by a lead–acid battery and 12–240 v inverter. Tapes and batteries could be changed without causing any disturbance at the nest. Continuous recordings on to 4-hour tapes were made which could be analysed together with the parallel observations on the foraging areas used by the adults (Sitters, 1991). As an example of the quality of data obtained, Fig. 6.2 shows the percentage of feeding visits when prey brought to the nest was identified as invertebrate or cereal grain, plotted against date. The usual pattern is for small young to be fed mainly on invertebrates, but for the diet to shift towards a higher proportion of cereal grains (mainly barley) as they become older. In Fig. 6.2 it is clear that this pattern has been disrupted on the days with rain (July 7, 8 and 11–13), when a much higher fraction of cereal grains was brought. It was also found that on these wet days the adults spent more minutes per day brooding the young. Presumably the weather-induced switch to cereal grains was forced on the adults as invertebrates became difficult to obtain, and the time available for foraging decreased.

One of the key causes of breeding failures in cirl buntings is starvation of the young in the nest during periods of adverse weather (Evans, 1991), and these video data provide a clue to what may be happening. In areas where barley grains or other large seeds are unavailable, or early in the season before they are ripe, the cirl buntings may not be able to find suitable

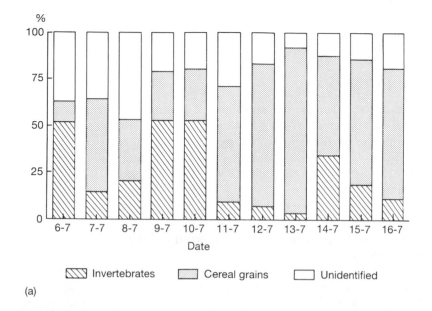

(a)

(b)

Figure 6.2 The daily percentage of feeding visits (a) when adult cirl buntings were seen on the video images (b) to be bringing cereal grain or invertebrates to the nest. Weather was wet on 7, 8 and 11–13 July.

alternative foods to bring to the young when invertebrates are difficult to obtain. At one of the study sites birds were collecting invertebrates from within 50 m of the nest, but were flying over 250 m to the nearest barley field to collect grain (Sitters, 1991).

6.4 DISCUSSION

Most of the video camera applications described here exploit the advantages, such as remote filming and the large image storage capacity of video cassettes, while finding ways around the key problems of cost and the provision of an adequate power supply. In the majority of applications the low light capability of the cameras was a bonus, but not the key reason for moving to video.

In future a wider use of video can be predicted, with camcorders coming into more widespread use. The higher resolution offered by some of the new formats will be particularly useful, especially if they become available for time-lapse systems.

One of the problems that has become obvious with our work is the time taken to analyse the masses of data collected by video systems. Automated analysis systems are already available for simple setups, but we can expect these to become more sophisticated in future. In the security market, systems are already available which trigger alarms when changes occur in predefined areas of a video image. Although these are still expensive, they could potentially provide a very neat way of recording video data only when interesting events were taking place in the subject area.

REFERENCES

Batten, L.A., Bibby, C.J., Clement, P. *et al.* (1990). *Red Data Birds in Britain*, Poyser, London.

Davies, M. and Campey, R. (1987). *RSPB Operation Woodpecker*, RSPB, Sandy.

Davis, P.E. and Newton, I. (1981). Population and breeding of Red Kites in Wales over a 30-year period. *Journal of Animal Ecology*, **50**, 759–772.

Dyer, P.K. and Hill, G.J.E. (1991). A solution to the problem of determining the occupancy status of Wedge-tailed Shearwater *Puffinus pacificus* burrows. *Emu*, **91**, 20–25.

Evans, A.D. (1991). *Cirl Buntings: Report on Research Project September 1988–May 1991*, RSPB, Sandy.

Evans, A.D. (1992). The numbers and distribution of Cirl Bunting *Emberiza cirlus* breeding in Britain in 1989. *Bird Study*, **39**, 17–22.

Haftorn, S. (1972). Closed-circuit television and datalogger, new tools to ornithology. *Sterna*, **11**, 243–252.

Haftorn, S. (1978a). Behaviour of the Goldcrest *Regulus regulus* during the act of egg laying, as observed on closed-circuit TV. *Cinclus*, **1**, 55–57.

Haftorn, S. (1978b). The behaviour of young Goldcrests *Regulus regulus* on leaving the nest, as observed on closed-circuit TV. *Cinclus*, **1**, 48–54.

Haftorn, S. (1978c). Cooperation between the male and female Goldcrest *Regulus regulus* when rearing overlapping double broods. *Ornis Scand.*, **9**, 124–129.

Haftorn, S. (1984). The behaviour of an incubating female Coal Tit *Parus ater* in relation to experimental manipulation of nest temperature. *Fauna norv. Ser. C, Cinclus*, **7**, 12–20.

Hussell, D.T.J. (1974). Photographic records of predation at Lapland Longspur and Snow Bunting nests. *Canadian Field Naturalist*, **88**, 503–506.

Lovegrove, R. (1990). *The Kite's Tale*, RSPB, Sandy.

Lovegrove, R., Elliot, G. and Smith, K. (1990). The Red Kite in Britain. *RSPB Conservation Review*, **4**, 15–21.

Major, R.E. (1991). Identification of nest predators by photography, dummy eggs and adhesive tape. *Auk*, **108**, 190–195.

Mudge, G.P., Aspinall, S.J. and Crooke, C.H. (1987). A photographic study of seabird attendance at Moray Firth colonies outside the breeding season. *Bird Study*, **34**, 28–36.

Newton, I. (1979). *Population Ecology of Raptors*, Poyser, Calton.

Perrins, C.M. (1979). *British Tits*, Collins, London.

Picman, J. (1987). An inexpensive camera set-up for the study of egg predation at artificial nests. *Journal of Field Ornithology*, **58**, 372–382.

Savidge, J.A. and Seibert, T.F. (1988). An infrared trigger and camera to identify predators at artificial nests. *Journal of Wildlife Management*, **52**, 291–294.

Sitters, H.P. (1991). A study of foraging behaviour and parental care in the Cirl Bunting *Emberiza cirlus*. MSc Thesis, University of Aberdeen.

Smith, K.W. (1987). *Nest Attendance Monitoring in Buzzard and Red Kite*, RSPB, Sandy.

Smith, K.W. (1988). *Report of Red Kite Research in 1988*, RSPB, Sandy.

Smith, K.W., Doody, D., Cartmel, S. and Prior, P. (1991). *Report of RSPB Red Kite Research in 1990 and 1991*, RSPB, Sandy.

Tommeraas, P.J. (1989). A time-lapse nest study of a pair of Gyrfalcons *Falco rusticolus* from their arrival at the nesting ledge to the completion of egg-laying. *Fauna Series C, Cinclus*, **12**, 52–63.

Walters-Davies, P. and Davis, P.E. (1973). The ecology and conservation of the Red Kite in Wales. *British Birds*, **66**, 183–224, 241–270.

Wisniewski, L. (1983). Remote video monitoring of eagles and falcons, in *Bald Eagle Restoration* (ed. T.N. Ingram). Eagle Valley Environmentalists, Apple River, Illinois, pp. 58–61.

7

Farm animals

C. Sherwin

7.1 INTRODUCTION

Compared with most other species discussed in this book, farm animals offer several advantages when their behaviour is to be video recorded. They are physically large, which means that expensive macro-equipment is not required; they can be easily located; and there are often many individuals in one location, allowing rapid collection of large amounts of data. These benefits, in addition to recent technological advances and decreases in the prices of equipment, might lead one to expect increasing usage of video recording in farm animal ethology. However, a survey of the most comprehensive scientific journal of farm animal behaviour, *Applied Animal Behaviour Science*, has revealed this to be erroneous. The number of studies using video to record the behaviour of farm animals has increased considerably (Fig. 7.1a), but when this is expressed as a proportion of total studies on farm animals (Fig. 7.1b) it appears to be constant. This might indicate that current video technology has little new to offer farm animal ethology, or that workers in this discipline are remaining ignorant of the possible techniques and advantages of video recording.

Farm animal ethologists who regularly use video-cassette recording (VCR) develop techniques to suit their own purposes. These techniques are generally simple, but their development by an individual researcher can be time-consuming, frustrating and fraught with losses of data. This chapter is most likely to benefit persons at greatest risk of making such mistakes, i.e. researchers using VCR for the first time, or on a species unfamiliar to them. Subjects are presented in the order they are likely to be contemplated when planning and executing a study. These include techniques of marking, camera use and recording, playback and data extraction, and general

Video Techniques in Animal Ecology and Behaviour. Edited by Stephen D. Wratten. Published in 1993 by Chapman & Hall, London. ISBN 0 412 46640 6

Farm animals

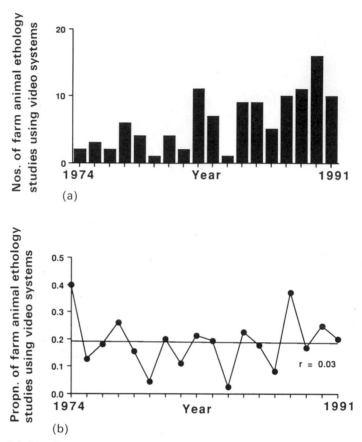

Figure 7.1 (a) The numbers of studies published in *Applied Animal Behaviour Science* which used video systems to record the behaviour of farm animals. (b) The proportion of farm animal ethology studies published in *Applied Animal Behaviour Science* which used a video recording system.

complications. Prior to these, there is a brief discussion of how comparable are data collected by direct observation or VCR, and an overview of studies that have used VCR to record farm animal behaviour.

7.2 HOW COMPARABLE ARE DATA GAINED BY DIRECT OBSERVATION AND VIDEO RECORDING?

Although video recording generates an accurate record of behaviour, the method by which data are extracted from the video tape can cause discrepancies between results gained using this technique and those from direct observations.

McDonnell and Diehl (1990) compared data collected by a single observer during direct observations, and from video tapes of the same behaviour (sexual behaviour of horses). Whereas data gained using the two methods were always significantly correlated, the relationship was rarely perfect. In fact, the correlations were as low as 0.48. Long-duration activities were quantified similarly by the two methods, but short-duration or frequency counts were less consistent. Similarly, Arnold-Meeks and McGlone (1986) reported that viewing tapes at $4 \times$ real time provided an accurate record of attack, feeding, drinking, lying, moving and standing; when viewed at $24 \times$ real time, discrepancies occurred for drinking and feeding. It has been noted that some behaviours, especially slow small movements, are less easily observed during video playback (Stamp Dawkins, 1982; Rushen, 1984; Boe, 1990). Such behaviours should therefore be quantified cautiously when using VCR, as data can unknowingly be lost.

Although direct observation and video recording might yield dissimilar results, this does not necessarily mean the video results are inaccurate: quite the opposite might be true. Video records have the distinct advantage that they can be replayed many times until accuracy is assured. The same cannot be said for direct observations.

7.3 AN OVERVIEW OF FARM ANIMAL BEHAVIOUR STUDIES USING VCR

VCR is usually used to record animal behaviour for one or more of the following reasons: first, it is often convenient to substitute a camera for an observer, thus allowing the observer to use his or her time more flexibly; secondly, VCR is sometimes the only method of obtaining the required data: some behaviours occur so rapidly or frequently that it is impossible for a human to accurately observe and record in real time; thirdly, some behaviours are influenced by the presence of an observer, making indirect observation essential; and fourthly, VCR is sometimes used as a means of validating data gained by direct observation. The following section reviews some of the behavioural studies in which VCR has been used to record the behaviour of farm animals for these reasons.

7.3.1 Video recording as a 'convenience' technique

If direct observations are conducted for any great length of time, observer fatigue can threaten the validity of data. Altmann (1974) stated that fatigue can limit direct observations to a 15-minute period each hour. In addition, repeated observations or those requiring several observers can be extremely disruptive to other commitments. These problems can be nullified by substituting a VCR system for the observer. The following sections show the types of behaviours which might be recorded by VCR for 'convenience' reasons.

Farm animals

Infrequent behaviours

If the behaviour is infrequent or unpredictable in its timing, it requires a considerable time investment to collect data by direct observation. Studies which have used VCR to avoid this include examinations of stereotypic activities (Redbo, 1990), dust-bathing by hens (Van Liere *et al.*, 1990), masturbation by cattle (Houpt and Wollney, 1989), selection of resting sites by pigs (Fraser, 1985), cattle approaches to an electric fence (McDonald *et al.*, 1981), parturition in goats (Lickliter, 1985) and egg-laying behaviour (Rietveld-Piepers *et al.*, 1985).

Long-duration behaviours

Many behaviours can take several hours to complete, or must be monitored over several days or weeks. By using VCR this data can be collected in the absence of an observer. Such studies include thermoregulation by sheep (Sherwin and Johnson, 1989; Boe, 1990), activity levels of calves (Weiguo and Phillips, 1991), roosting of hens (Muiruri *et al.*, 1990), spatial use of the enclosure by cattle (Ganskopp *et al.*, 1991), time budgets of bulls (Houpt and Wollney, 1989) and the development of nursing behaviour in pigs (Lewis and Hurnik, 1985).

Behaviour dependent on the time of day

If a behaviour is influenced by the time of day, then observations are required over the entire day or during representative periods. Behaviours which are influenced in this manner and have been recorded using VCR include stereotypies in heifers (Redbo, 1990), shade-seeking by sheep (Sherwin and Johnson, 1987), dust-bathing by quail (Schein and Statkiewicz, 1983) and bar-biting by sows (Jensen, 1988).

When conditions are uncomfortable for humans

Some farm animals, notably pigs and poultry, are often housed in environments which are very unpleasant for humans to work in. The major problems are excessive noise, ammonia, dust and (less frequently) extreme temperatures. Using VCR is most convenient – sometimes necessary – under these conditions. Standard video equipment is relatively insensitive to levels normally encountered in animal husbandry systems, though it should be protected from dust and water (see below). Using VCR can also prevent ethologists having to place themselves in potentially dangerous situations, such as travelling in a lorry surrounded by several tonnes of excited cattle (Tarrant and Kenny, 1986; Lambooy and Hulsegge, 1988).

7.3.2 Video recording as an essential technique

Some behaviours occur too rapidly or frequently for human senses to reliably detect, or involve many animals such that quantification requires more observers than are available. Such behaviours can be recorded and played back repeatedly, in slow motion if necessary.

Extremely rapid or frequent behaviours

Rapid or frequent behaviours analysed by VCR include hen 'comfort' behaviours (e.g. tail-wagging, wing movements, body and head shaking) (Nicol, 1986), feeding by hens (Gentle *et al.*, 1982; Van Rooijen, 1991) and drinking by broiler chicks (Ross and Hurnik, 1983).

Highly complex behaviours

Some behaviours are comprised of many small movements which are too complex to examine by real-time direct observations. These include teat-seeking (Rohde and Gonyou, 1987; Rohde Parfet and Gonyou, 1990), head-tracking movements of chicks during feeding (Rogers and Andrew, 1989) and responses of hens following a period of spatial restriction (Nicol, 1987).

Social behaviours

Social behaviours are frequently subtle and complex. By using VCR and replaying the tape several times, the required data can be gradually extracted. Such studies include social orientation in hens (McCort and Graves, 1978; Mankovich and Banks, 1982), agonistic interactions in pigs (Jensen, 1980, 1982; Fraser *et al.*, 1991) and hens (Stamp Dawkins, 1982), priority use of the feeder by pigs (Hunter, 1988) and sheep (Sherwin, 1990a), leadership in cattle (Boissy and Bouissou, 1988) and sheep (Squires and Daws, 1975; Hutson, 1980; Sherwin, 1990a), deer social behaviour (Walton and Hosey, 1984) and the responses of sheep to trained dogs (Anderson *et al.*, 1988).

Unpredictable behaviours

If it is impossible to predict a behavioural response, it can be advantageous to video record these for later analysis. An example of this is a study by Sherwin and Nicol (1991) on the effect of transport on broiler hens. It was hypothesized that transport would influence the awareness of hens, but it was not possible to predict precisely how this might be exhibited. An auditory-conditioned stimulus was used to signal the onset of a mildly aversive stimulus (spraying with water). By video-recording responses after the conditioned stimulus, it was possible to extract data on vocalizations, locomotion, defaecation, posture changes and head-shaking to look for differences in awareness between transported and non-transported birds. Other studies where video recording has been used, apparently because responses were difficult to predict, include visual cues in the mutual recognition of ewes and lambs (Alexander, 1977), social deprivation of lambs (Zito *et al.*, 1977), transport of heifers (Lambooy and Hulsegge, 1988), oestrous behaviour of sows (Bressers *et al.*, 1991) and the preovulatory behaviour of cattle (Esslemont *et al.*, 1980).

Standardization of behaviour

An interesting and relatively recent use of VCR is to standardize the behaviour of 'demonstrator' animals when analysing behavioural responses. This is done by playing a video recording of a 'demonstrator' to an 'observer' and recording the responses of the latter. It has been shown that several species, including sheep (Franklin and Hutson, 1982) and hens (Evans and Marler, 1991), respond to recorded images of conspecifics as if they were real animals. This method is currently being used by several workers, including Dr Christine Nicol at the University of Bristol. She is showing hens video recordings of conspecifics being handled and attempting to determine if the observers find the distress of another bird aversive.

7.3.3 Video recording as a technique to eliminate observer effects

Human presence, especially when associated with feed or handling, can have a considerable effect on the behaviour of farm animals. Hides can be used when taking direct observations, although many farm animal ethologists prefer to sit quietly in view of the animals for several minutes, allowing habituation prior to observations. Despite the fundamental importance of collecting data from undisturbed animals, few workers have quantified the effects of observer presence. Studies which have used VCR to overcome this problem include those on the mating of geese (Gillette, 1977), stereotypies in sows (Rushen, 1984, 1985), social interactions between calves (Canali *et al.*, 1986) and open-field behaviour of rabbits (Meijsser *et al.*, 1989). It is likely that many other behaviours are influenced by observer presence, and studies on these would benefit by the use of VCR.

7.3.4 Video recording as a technique to validate direct observations

Video records of farm animal behaviour have been made as a means to validate or check data collected by direct observation (Shillito and Alexander, 1975; Arnold, 1982; Arnold-Meeks and McGlone, 1986).

7.4 IDENTIFICATION AND MARKING TECHNIQUES

In ethological studies it is frequently necessary to identify individuals within a group. This can be difficult when observing farm animal species, since these are often remarkably uniform. Genetic selection and husbandry practices such as horn removal can exacerbate this difficulty by eliminating obviously variable features. Marking animals in a manner which enables individual recognition might seem a relatively simple task, but in practice it can be one of the greatest hurdles to overcome when collecting behavioural data using VCR. Several factors should be considered when deciding on the identification method to be used.

1. The method must be appropriate for the camera position and the morphology and behaviour of the animal. For example, leg bands on hens cannot be filmed by an overhead camera, and numbered ear-tags are too small if there is any great distance between the camera and the animal.
2. The duration of observations should be considered. Observations of farm animals usually take place over a few days or weeks, for which short-term marks are used, e.g. the application of paint, dye, bleach, crayon, numbered saddles, collars or number tags. These methods are relatively non-invasive, convenient and inexpensive. Observations over a longer duration require more long-term markings, i.e. ear-tags, branding, tattooing, mutilations. These methods frequently compromise the welfare of the animals, are more difficult to apply and are generally more expensive. It can be advantageous to use both short- and long-term marks. If the short-term mark is lost or becomes illegible, the long-term mark, although perhaps not observable on video recordings, can be used as a back-up for accurate remarking.
3. The marking method should not disturb the animals' behaviour. Placing ear-tags, leg bands, collars, or even simply painting numbers on animals, often causes a temporary increase in investigative behaviour. In pigs, rubbing and chewing of newly attached ear-tags can become so intense that lesions occur, necessitating veterinary treatment (Sherwin, 1990b).
4. Toxicity of the mark should be considered. Most commercially available animal spray paints are approved for use on food animals, but application of other chemicals such as dyes, bleach (Price *et al.*, 1985) or gloss paint (Phillips and Leaver, 1986) should be considered carefully from the animal welfare and human consumption perspectives.
5. Keeping the identification scheme simple reduces the probability of misidentifying individuals. Colour schemes are particularly problematic when using VCR, since discriminating between colours becomes difficult when recording in dim conditions. Colours should be used which are widely dispersed on the visual spectrum, e.g. red, yellow and blue. Some numerals (1 and 7, 0 and 8, 3 and 8) can be easily confused, especially

when applied to an irregular surface such as a long-wooled sheep. Obscuration of one digit when numbers greater than 9 are used can also easily cause misidentification, e.g. '16' or '26' can be misread as '6'.

Methods of marking farm animals for identification during studies using VCR are described below.

Poultry: when using light-coloured or white birds, identification stripes can be applied along the back with paint (Ramos and Craig, 1988) or felt pen (Rietveld-Piepers *et al.*, 1985). These need remarking approximately once a month. Successful marking of darker hens poses more of a problem. Stamp Dawkins (1982) used natural variation in plumage to distinguish individuals. Leg-bands can be used, although identifying colours from video tapes can be difficult. Other identification methods are coloured/numbered bands which fit around the wing, or – less satisfactorily from a welfare point of view – pierce the wing. Painting numerals or complex symbols on poultry is inadvisable, since the mark can be easily distorted during wing-stretching or feather ruffling. Workman and Andrew (1989) avoided this by 'colour-marking' the crowns of chicks. The author is currently investigating the use of fat-soluble dyes (e.g. Sudan IV) for colouring wattles and combs.

Sheep: characters are easily painted on the wool of sheep (Fig. 7.2) and can remain identifiable for many weeks, although this is largely determined by climatic conditions and wool growth. Sherwin (1990a) glued

Figure 7.2 A sheep with a painted number on the flank and carrying a numbered saddle.

cloth squares with painted numbers to the backs of sheep (Fig. 7.3), and similarly, Tilbrook *et al.* (1987) identified individuals by fluorescent characters painted on saddles strapped to the sides of ewes. Horns can be painted with coloured stripes (Fowler and Jenkins, 1976) but this is obviously restricted to horned animals. Numbered/coloured ear-tags can be used, but these are usually too small to be practicable for video recording sheep in fields.

Goats: painted characters, identifying collars (Lickliter, 1982) and ear-tags can all be used to identify goats, but attachment of saddles, etc. should probably be avoided due to their infamous chewing habits.

Pigs: most farmed pigs have relatively hairless bodies and characters are easily applied with commercial spray-paints or crayons. Pigs frequently huddle and rub against objects, thereby smearing and removing such marks. Jensen *et al.* (1986) commented that painted numbers on older pigs remain

Figure 7.3 Sheep with numbered cloths glued to their wool.

legible for approximately 3 weeks; with younger pigs this is reduced to 1 week. It is advisable when applying paint or crayon to group-housed pigs (or indeed any group-housed farm animal) to mark only one or two individuals at a time, or, after marking, to release the animals to a large enclosure where they will not congregate into a tight bunch and smear the paint. Numbered ear-tags and collars can be used, but both can be removed by the pig or its penmates by persistent rubbing or chewing. Collars must be adjusted frequently (every 2–3 days) if the animals are in a phase of rapid growth. Tattoos and ear-notching might be used, but should be avoided for welfare reasons.

Cattle: natural variations in the coat and/or horns of cattle can often be used to identify individuals: a set of identification photographs aids this method. Characters can be painted or dyed into the coat, or identifying collars and ear-tags used. Again, horns can be painted with coloured stripes. Edmunds (1975) described a method whereby a 'box-tag' attached to a small saddle of cloth is glued to the withers. There is no obvious reason why this method could not be used for several other species. Branding (freeze and hot-iron) and ear-notching might be used but again, welfare considerations should preclude these.

Horses and donkeys: farm horses have often been branded by previous owners, and although these marks are usually deliberately small, they might be adequate for video recording. Horses are quite variable in appearance (except thoroughbreds), otherwise the methods described for cattle are usually suitable.

Deer: numbered ear-tags and collars have been used to individually identify deer. Natural variability in coat and antlers offers an alternative and relatively long-term form of identification.

7.5 CAMERA AND RECORDING TECHNIQUES

There is frequently no single position from which a static video camera can record the entire observation site. In addition, structures such as feed troughs cause 'blind spots'. Appropriate data analysis can overcome this problem, e.g. by expressing data as a proportion of all visible animals. Alternatively, the entire enclosure can be filmed by using mirrors or a number of appropriately positioned cameras (Friend and Polan, 1978). Multiplexing between recordings from several cameras (Lewis and Hurnik, 1985) is possible, or the recorded image can be a combination of several simultaneous inputs. This technique can also be used to record animals in several pens (Muiruri *et al.*, 1990; de Passille and Robert, 1989). Motorized pan-heads which rotate the camera can be used. These are manually activated at a site remote from the observations, or programmed to automatically rotate at regular intervals.

If an accurate analysis of three-dimensional behaviour is required, it is necessary to use one camera in the horizontal plane and one in the vertical (Rogers and Andrew, 1989).

Low-light (Houpt and Wollney, 1989) or infra-red cameras (Jensen, 1988; Lambooy and Hulsegge, 1988; Boe, 1991; Meunier-Salaun *et al.*, 1991) have been used to video record the behaviour of farm animals in dim conditions. Complications associated with these cameras are discussed below. Heat-sensitive cameras are now available which record the temperature distribution of the objects they film. Such cameras could enable various important studies to be conducted on thermoregulatory behaviour or welfare. The high cost of this equipment has probably prevented such uses to date.

If observer presence is likely to disturb the animals' behaviour, observations can be made remotely by using the VCR camera and monitor as a closed-circuit monitoring system, as seen in security applications (Gallup and Suarez, 1980; Hargreaves and Hutson, 1990).

A highly flexible technique of VCR is the timing of data collection. Time-lapse is frequently used, whereby the system is programmed to record a limited number of frames in each time unit (e.g. 4 fps). This conserves space on the tape and enables data to be collected over a longer period of time. Similarly, the VCR system can be programmed to start recording at a specified time, or to turn on and off for set durations (see Poysa (1991) for further discussion of time-sampling). There are often periods within observation sessions when a video record is not required, e.g. when one animal has to be removed from the test room and another brought in. During these periods it is tempting to turn off the recorder and save space on the tape. However, it is remarkably easy to forget to turn on the recorder when observations recommence, thus causing disastrous data losses. Even when observations are to be conducted over several hours, it is wise to activate the VCR system at the beginning of the session and leave it recording until the very end.

7.6 PLAYBACK AND DATA EXTRACTION TECHNIQUES

During playback of video tapes several techniques can be used to facilitate observation of the images and data extraction.

7.6.1 Adjustment of the monitor

One method of 'improving' video-recorded images is simply to adjust the monitor until the various qualities are optimal. If the recording is in colour, it is sometimes beneficial to override this and watch the playback in black and white. Presumably this has its enhancing effect by removing distracting colourful details. Similarly, adjusting the contrast and/or brightness to extreme levels can make it considerably easier to see details, such as heads over a feed trough. Although easily achieved, adjusting the monitor can substantially reduce eye strain and avoid observer fatigue.

7.6.2 High-speed playback

By using the high-speed playback mode it is possible to rapidly scan through a recording until the desired behaviour occurs, then observe this at normal speed to obtain the required details. Alternatively, the tape can be played continuously at the faster speed. Arnold-Meeks and McGlone (1986) compared data collected during direct observations, normal-speed and high-speed playback. Some discrepancies were noted (see above), indicating that if a high degree of accuracy is required, data extraction from high-speed playback should be avoided.

7.6.3 Frame-by-frame analysis

Frame-by-frame analysis has been used to analyse patterns of agonistic interactions between sows (Jensen, 1980, 1982), head movements, social behaviours and the actions involved in lying (Jensen *et al.*, 1986). Single-frame images from video recordings are often of inferior quality, and other formats such as 8 mm cine film are better suited to this technique.

7.6.4 Drawing overlays

Using recordings from a video camera positioned overhead, Rogers and Andrew (1989) measured the lateral head movements of chicks during orientation and feeding. During playback of the tapes, a sheet of transparent celluloid was rolled at a constant speed over the monitor screen. A pen positioned over the image of the chick's comb traced head movements on to the celluloid sheet.

7.6.5 Event recorders

Many researchers have quantified farm animal behaviour by using an event recorder during playback (e.g. Rushen, 1984; McDonnell and Diehl, 1990).

7.6.6 Magnification of recorded images

It is possible to project VCR images on to a large screen (Stuth *et al.*, 1987). This technique facilitates the identification of animals and increases the distance between individuals, thus improving accuracy when measuring spacing behaviours.

7.6.7 Digitization of position/orientation

Several workers have reported using computerized digitizers to record small or frequent movements (McCort and Graves, 1978; Mankovich and Banks,

1982; Rohde and Gonyou, 1987; Rohde Parfet and Gonyou, 1990). The sensor of a digitizer is mounted over the video monitor (or the image is projected on to the sensor) and then during playback the stylus is touched to the position of interest, e.g. the snout during suckling. This generates spatial coordinates and allows the position of an animal or a part of its body to be plotted very accurately and repeatedly with little effort.

7.7 GENERAL COMPLICATIONS

7.7.1 Observer fatigue

Although using VCR can improve time management and will sometimes save time compared to direct observations, it should be remembered that the tapes must be watched at some time to extract the data. There is a tendency to store recordings and delay watching the tapes until the last possible moment. Frequently, this leads to the researcher being forced to watch a monitor for many hours within a short space of time, thus developing considerable fatigue – precisely one of the problems that VCR is supposed to eliminate!

7.7.2 Limited field of view

Sometimes incidents occur 'off-screen' (e.g. humans approaching the enclosure) which disturb the behaviour being quantified. If no observer is present during recording, it can be impossible to account for such disturbances when interpreting the data. This problem can also arise during direct observations, but it is more likely to be noticed since humans can visually scan a wider area than a static video camera. The field of view can be increased by using wide-angle lenses or motorized pan-heads, but this specialist equipment is usually expensive and image clarity is sometimes sacrificed. Extraneous noises which disturb behaviour will remain unnoticed when tapes are played back at high speed, since the audio signal is usually deactivated.

7.7.3 Protection of equipment

Farm animals will forcefully investigate and damage objects placed in their enclosure. Protection of video equipment is therefore best achieved by positioning it beyond the animals' reach, usually outside the enclosure, or raised several metres. This has the added advantage of increasing the field of view. Video equipment should also be protected from the climate. If recordings are to be made in a dusty environment (notably pig and poultry sheds) a plastic bag or sheet over the system is usually sufficient, with a hole cut in the sheet, the camera lens placed through and sealed with an elastic band. This also protects against water overspray when the enclosures are being cleaned, or mist in a partially open building such as a hay shed. For outdoor studies, extreme ambient temperatures and heavy rain should be considered. Most VCR systems are unaffected by the air temperatures farm animals normally experience, although placing equipment in direct sunlight on a hot day can cause malfunctions; some form of shade is advisable. To satisfactorily protect equipment from both climate and animals, a sturdy hide or shed is best. This must be secure enough to prevent damage by animals that rub or lean on such structures. One method of achieving high stability is to secure the shed to a heavy base such as a palette (Fig. 7.4). This has the additional benefit that the structure is then relatively portable, as it can be moved by a fork-lift or tractor.

Figure 7.4 An observation shed securely attached to a wooden palette (the top one) thus affording stability and portability.

7.7.4 A power source is required

Many VCR systems require a mains power source, although purpose-designed portable units can be powered by either batteries or mains. A mains power extension cable can be used, but voltage drop means there is a finite limit to the distance over which these can be used. While making video recordings of sheep, Sherwin and Johnson (1989) used an extension cable over 400 m long. It is important to adequately protect the entire cable from chewing and other accidental damage, since it carries a potentially lethal current. Batteries allow total portability, but they are expensive and have a lifespan too short for many ethological studies.

7.7.5 Video recording requires light

Most standard video cameras cannot record acceptable images below 7 lux (approximately the level experienced in an average-sized office lit by one 40 W incandescent bulb). Supplementary lighting is therefore required if recordings are to be made at night. This has been achieved by leaving lights on continuously (Arey *et al.*, 1991; Geers *et al.*, 1986; Houpt and Wollney, 1989; Widowski and Curtis, 1990), having a 'bright' and a 'dim' period (Fraser *et al.*, 1991), or by using continuous low-intensity coloured light (Feddes *et al.*, 1989; Muiruri *et al.*, 1990). However, the effects of such lighting on behaviour have received little attention. Reports of observations under such conditions sometimes state that the behaviour appeared unaffected by the lights, though quantified evidence is rarely offered. Short-term illumination of animals at night using a torch or spotlight appears to have little effect on the behaviour of cows (Phillips and Leaver, 1986; Ganskopp *et al.*, 1991), sheep (Lynch *et al.*, 1980), pigs (Petersen *et al.*, 1990) and hens (Appleby and Smith, 1991), but long-term changes in lighting have been shown to influence behaviour. Weiguo and Phillips (1991) reported that supplementary light reduced daily activity levels in calves, and Dannenmann *et al.* (1985) found that greater light intensity increased playing and licking, but decreased the frequency of resting. Lammers and de Lange (1986) reported that greater activity levels of pigs were associated with the light phase, and Feddes *et al.* (1989) showed that feeding activity increased around the onset and termination of the light phase. Systematic studies on the effects of supplementary light used to allow video recording are required. Until then, we should be more cautious in our interpretation and acceptance of data gained by this method.

Similarly, infra-red cameras usually use a far/infra-red light source to 'illuminate' the animals. Although this technique is widely used, there have been few studies on whether these lights influence behaviour. Mistry and McCracken (1990), in an experiment using Mexican free-tailed bats, showed that infra-red and far-red light did not influence escape behaviour. However, this does not mean that the bats could not perceive these frequencies and, moreover, the possibility remains that other behaviours in different species might be affected.

7.8 THE FUTURE OF VCR IN FARM ANIMAL ETHOLOGY

There are several (potential) VCR techniques that have not yet been applied to farm animal ethology, or that remain to be fully explored. Some of these require further technological development, whereas others could be achieved with currently available equipment.

Image analysis systems are available (e.g. Columbus Instruments, USA) which automatically record behaviours such as spatial use of the cage, proximity to neighbours, periods of activity/inactivity, etc. These systems, which rely on high contrast between the animal and the background, are usually used for laboratory animal studies. The behaviour of farm animals could be analysed using this technique by painting the floor of the enclosure an appropriate colour.

VCR combined with image analysis systems could be used in several practical aspects of husbandry. By monitoring the spatial use of the enclosure, oestrus could be determined by assessing when animals spend more time near members of the opposite sex. There is a strong correlation between the weight of a pig and the area of its plan view without head and neck (Schofield and Marchant, 1990). Using this information, an integrated VCR and image analysis system could inform the farmer when to adjust feed amounts, pen size, group size, etc. Palm and fingerprint image analysis are available for human security systems: similar technology could be applied to aid identification of animals by using their nose-prints.

New methods of slaughter or stunning, e.g. CO_2, could be examined by movement monitoring and thermal imaging to determine when death is imminent. Other scientific studies (e.g. the welfare of poultry during transport, use of enclosures during thermoregulatory behaviour) could also benefit from such systems.

Presenting video-recorded images to animals is a technique which deserves considerable attention. Standard husbandry practices might be improved, e.g. increasing the speed of animal movement in raceways at abattoirs, lairage etc. by projecting images of conspecifics on a screen at the end of the raceway (Franklin and Hutson, 1982). The use of prerecorded images might also be used to improve animal welfare by enriching the environment, or manipulating behaviour to promote positive activities while reducing deleterious ones.

REFERENCES

Alexander, G. (1977). Role of auditory and visual cues in mutual recognition between ewes and lambs in Merino sheep. *Applied Animal Ethology*, **3**, 65–81.

Altmann, J. (1974). Observational study of behaviour: sampling methods. *Behaviour*, **49**, 227–267.

Anderson, D.M., Hulet, C.V., Shupe, W.L. *et al.* (1988). Response of bonded and non-bonded sheep to the approach of a trained border collie. *Applied Animal Behaviour Science*, **21**, 251–257.

Appleby, M.C. and Smith, S.F. (1991). Design of nest-boxes for laying cages. *British Poultry Science*, **32**, 667–678.

Arey, D.S., Petchey, A.M. and Fowler. V.R. (1991). The preparturient behaviour of sows in enriched pens and the effect of pre-formed nests. *Applied Animal Behaviour Science*, **31**, 61–68.

Arnold, G.W. (1982). Ethogram of agonistic behaviour for thoroughbred horses. *Applied Animal Behaviour Science*, **8**, 5–25.

Arnold-Meeks, C. and McGlone, J.J. (1986). Validating techniques to sample behaviour of confined, young pigs. *Applied Animal Behaviour Science*, **16**, 149–155.

Boe, K. (1990). Thermoregulatory behaviour of sheep housed in insulated and uninsulated buildings. *Applied Animal Behaviour Science*, **27**, 243–252.

Boe, K. (1991). The process of weaning in pigs: when the sow decides. *Applied Animal Behaviour Science*, **30**, 47–59.

Boissy, A. and Bouissou, M.F. (1988). Effects of early handling on heifers' subsequent reactivity to humans and to unfamiliar situations. *Applied Animal Behaviour Science*, **20**, 259–273.

Bressers, H.P.M., Te Brake, J.H.A. and Noordhuizen, J.P.T.M. (1991). Oestrus detection in group-housed sows by analysis of data on visits to the boar. *Applied Animal Behaviour Science*, **31**, 183–193.

Canali, E., Verga, M., Montagna, M. and Baldi, A. (1986). Social interactions and induced behavioural reactions in milk-fed female calves. *Applied Animal Behaviour Science*, **16**, 207–215.

Dannenmann, K., Buchenauer, D. and Fliegner, H. (1985). The behaviour of calves under four levels of lighting. *Applied Animal Behaviour Science*, **13**, 243–258.

de Passille, A.M.B. and Robert, S. (1989). Behaviour of lactating sows: influence of stage of lactation and husbandry practices at weaning. *Applied Animal Behaviour Science*, **23**, 315–329.

Edmunds, J. (1975). A method of animal identification in studies on multisuckling behaviour. *Applied Animal Ethology*, **2**, 93–95.

Esslemont, R.J., Glencross, R.G., Bryant, M.J. and Pope, G.S. (1980). A quantitative study of pre-ovulatory behaviour in cattle (British Fresian heifers). *Applied Animal Ethology*, **6**, 1–17.

Evans, C.S. and Marler, P. (1991). On the use of video images as social stimuli in birds: audience effects on alarm calling. *Animal Behaviour*, **41**, 17–26.

Feddes, J.J.R., Young, B.A. and DeShazer, J.A. (1989). Influence of temperature and light on feeding behaviour of pigs. *Applied Animal Behaviour Science*, **23**, 215–222.

Fowler, D.G. and Jenkins, L.D. (1976). The effects of dominance and infertility of rams on reproductive performance. *Applied Animal Ethology*, **2**, 327–337.

Franklin, J.R. and Hutson, G.D. (1982). Experiments on attracting sheep to move along a laneway. III. Visual stimuli. *Applied Animal Ethology*, **8**, 457–478.

Fraser, D. (1985). Selection of bedded and unbedded areas by pigs in relation to environmental temperature and behaviour. *Applied Animal Behaviour Science*, **14**, 117–126.

Fraser, D., Phillips, P.A., Thompson, B.K. and Tennessen, T. (1991). Effect of straw on the behaviour of growing pigs. *Applied Animal Behaviour Science*, **30**, 307–318.

Friend, T.H. and Polan, C.E. (1978). Competitive order as a measure of social dominance in dairy cattle. *Applied Animal Ethology*, **4**, 61–70.

Gallup, G.G. and Suarez, S.D. (1980). An ethological analysis of open-field behaviour in chickens. *Animal Behaviour*, **28**, 368–378.

Ganskopp, D., Raleigh, R., Schott, M. and Bracken, T.D. (1991). Behaviour of cattle in pens exposed to $\pm 500\,kV$ DC transmission lines. *Applied Animal Behaviour Science*, **30**, 1–16.

Geers, R., Goedseels, V., Parduyns, G. and Vercruysse, G. (1986). The group postural behaviour of growing pigs in relation to air velocity, air and floor temperature. *Applied Animal Behaviour Science*, **16**, 353–362.

Gentle, M.J., Hughes, B.O. and Hubrecht, R.C. (1982). The effect of beak trimming on food intake, feeding behaviour and body weight in adult hens. *Applied Animal Ethology*, **8**, 147–159.

Gillette, D.D. (1977). Mating and other behavior of domestic geese. *Applied Animal Ethology*, **3**, 305–319.

Hargreaves, A.L. and Hutson, G.D. (1990). Changes in heart rate, plasma cortisol and haematocrit of sheep during a shearing procedure. *Applied Animal Behaviour Science*, **26**, 91–101.

Houpt, K.A. and Wollney, G. (1989). Frequency of masturbation and time budgets of dairy bulls used for semen production. *Applied Animal Behaviour Science*, **24**, 217–225.

Hunter, E.J. (1988). Social hierarchy and feeder access in a group of sows using a computerised feeder. *Applied Animal Behaviour Science*, **21**, 372–373.

Hutson, G.D. (1980). Visual field, restricted vision and sheep movement in laneways. *Applied Animal Ethology*, **6**, 175–187.

Jensen, P. (1980). An ethogram of social interaction patterns in group-housed dry sows. *Applied Animal Ethology*, **6**, 341–350.

Jensen, P. (1982). An analysis of agonistic interaction patterns in group-housed dry sows – aggression regulation through an 'avoidance order'. *Applied Animal Ethology*, **9**, 47–61.

Jensen, P. (1988). Diurnal rhythm of bar-biting in relation to other behaviour in pregnant sows. *Applied Animal Behaviour Science*, **21**, 337–346.

Jensen, P., Algers, B. and Ekesbo, I. (1986). Methods of Sampling and Analysis of Data in Farm Animal Ethology, Birkhauser Verlag, Basel, Sweden.

Lambooy, E. and Hulsegge, B. (1988). Long-distance transport of pregnant heifers by truck. *Applied Animal Behaviour Science*, **20**, 249–258.

Lammers, G.J. and de Lange, A. (1986). Pre- and post-farrowing behaviour in primiparous domesticated pigs. *Applied Animal Behaviour Science*, **15**, 31–43.

Lewis, N.J. and Hurnik, J.F. (1985). The development of nursing behaviour in swine. *Applied Animal Behaviour Science*, **14**, 225–232.

Lickliter, R.E. (1982). Effects of a post-partum separation on maternal responsiveness in primiparous and multiparous domestic goats. *Applied Animal Ethology*, **8**, 537–542.

Lickliter, R.E. (1985). Behaviour associated with parturition in the domestic goat. *Applied Animal Behaviour Science*, **13**, 335–345.

Lynch, J.J., Mottershead, B.E. and Alexander, G. (1980). Sheltering behaviour and lamb mortality amongst shorn Merino ewes lambing in paddocks with a restricted area of shelter or no shelter. *Applied Animal Ethology*, **6**, 163–174.

McCort, W.D. and Graves, H.B. (1978). A computer-assisted technique for processing spacing and orientation behavior. *Applied Animal Ethology*, **4**, 205–209.

McDonald, C.L., Beilharz, R.G. and McCutchan, J.C. (1981). Training cattle to control by electric fences. *Applied Animal Ethology*, **7**, 113–121.

McDonnell, S.M. and Diehl, N.K. (1990). Computer-assisted recording of live and videotaped horse behaviour: reliability studies. *Applied Animal Behaviour Science*, **27**, 1–7.

Mankovich, N.J. and Banks, E.M. (1982). An analysis of social orientation and the use of space in a flock of domestic fowl. *Applied Animal Ethology*, **9**, 177–193.

Meijsser, F.M., Kersten, A.M.P., Wiepkema, P.R. and Metz, J.H.M. (1989). An analysis of the open-field performance of sub-adult rabbits. *Applied Animal Behaviour Science*, **24**, 147–155.

Meunier-Salaun, M.C., Gort, F., Prunier, A. and Schouten, W.P.G. (1991). Behavioural patterns and progesterone, cortisol and prolactin levels around parturition in European (Large White) and Chinese (Meishan) sows. *Applied Animal Behaviour Science*, **31**, 43–59.

Mistry, S. and McCracken, G.F. (1990). Behavioural response of the Mexican free-tailed bat, *Tadarida brasiliensis mexicana*, to visible and infra-red light. *Animal Behaviour*, **39**, 598–599.

Muiruri, H.K., Harrison, P.C. and Gonyou, H.W. (1990). Preferences of hens for shape and size of roosts. *Applied Animal Behaviour Science*, **27**, 141–147.

Nicol, C.J. (1986). *A Study Of The Behavioural Needs Of Battery Housed Hens*, PhD thesis, University of Oxford.

Nicol, C.J. (1987). Behavioural responses of laying hens following a period of spatial restriction. *Animal Behaviour*, **35**, 1709–1719.

Petersen, V., Recen, B, and Vestergaard, K. (1990). Behaviour of sows and piglets during farrowing under free-range conditions. *Applied Animal Behaviour Science*, **26**, 169–179.

Phillips, C.J.C. and Leaver, J.D. (1986). The effect of forage supplementation on the behaviour of grazing dairy cows. *Applied Animal Behaviour Science*, **16**, 233–247.

Poysa, H. (1991). Measuring time budgets with instantaneous sampling: a cautionary note. *Animal Behaviour*, **42**, 317–318.

Price, E.O., Martinez, C.L. and Coe, B.L. (1985). The effects of twinning on mother–offspring behaviour in range beef cattle. *Applied Animal Behaviour Science*, **13**, 309–320.

Ramos, N.C. and Craig, J.V. (1988). Pre-laying behaviour of hens kept in single- or multiple-hen cages. *Applied Animal Behaviour Science*, **19**, 305–313.

Redbo, I. (1990). Changes in duration and frequency of stereotypies and their adjoining behaviours in heifers, before, during and after the grazing period. *Applied Animal Behaviour Science*, **26**, 57–67.

Rietveld-Piepers, B., Blokhuis, H.J. and Wiepkema, P.R. (1985). Egg-laying behaviour and nest-site selection of domestic hens kept in small floor-pens. *Applied Animal Behaviour Science*, **14**, 75–88.

Rogers, L.J. and Andrew, R.J. (1989). Frontal and lateral visual field use by chicks after treatment with testosterone. *Animal Behaviour*, **38**, 394–405.

Rohde, K.A. and Gonyou, H.W. (1987). Strategies of teat-seeking behavior in neonatal pigs. *Applied Animal Behaviour Science*, **19**, 57–72.

Rohde Parfet, K.A. and Gonyou, H.W. (1990). Directing the teat-seeking behaviour of newborn piglets: use of sloped floors and curved corners in the design of farrowing units. *Applied Animal Behaviour Science*, **25**, 71–84.

Ross, P.A. and Hurnik, J.F. (1983). Drinking behaviour of broiler chicks. *Applied Animal Ethology*, **11**, 25–31.

Rushen, J. (1984). Stereotyped behaviour, adjunctive drinking and the feeding periods of tethered sows. *Animal Behaviour*, **32**, 1059–1067.

Rushen, J.P. (1985). Stereotypies, aggression and the feeding schedules of tethered sows. *Applied Animal Behaviour Science*, **14**, 137–147.

Schein, M.W. and Statkiewicz, W.R. (1983). Satiation and cyclic performance of dustbathing by Japanese Quail (*Coturnix coturnix japonica*). *Applied Animal Ethology*, **10**, 375–383.

Schofield, C.P. and Marchant, J.A. (1990). *Imaging Systems for Livestock Health and Welfare*, British Society for Research in Agricultural Engineering, Association Members' Day, Sept 13th.

Sherwin, C.M. (1990a). Priority of access to limited feed, butting hierarchy and movement order in a large group of sheep. *Applied Animal Behaviour Science*, **25**, 9–24.

Sherwin, C.M. (1990b). Ear-tag chewing, ear rubbing and ear traumas in a small group of gilts after having electronic ear-tags attached. *Applied Animal Behaviour Science*, **28**, 247–254.

Sherwin, C.M. and Johnson, K.G. (1987). The influence of social factors on the use of shade by sheep. *Applied Animal Behaviour Science*, **18**, 143–155.

Sherwin, C.M. and Johnson, K.G. (1989). Variability in shading behaviour of sheep. *Australian Journal of Agriculture Research*, **40**, 177–185.

Sherwin, C.M. and Nicol, C.J. (1991). Head shaking as an indicator of the awareness state of broilers after fasting and transport, in *Applied Animal Behaviour: Past, Present and Future*, (eds M.C. Appleby, R.I. Horrell, J.C. Petherick and S.M. Rutter, Universities Federation for Animal Welfare, Herts, UK. p. 156.

Shillito, E. and Alexander, G. (1975). Mutual recognition amongst ewes and lambs of four breeds of sheep (*Ovis aries*). *Applied Animal Ethology*, **1**, 151–165.

Squires, V.R. and Daws, G.T. (1975). Leadership and dominance relationships in Merino and Border Leicester sheep. *Applied Animal Ethology*, **1**, 263–274.

Stamp Dawkins, M. (1982). Elusive concept of preferred group size in domestic hens. *Applied Animal Ethology*, **8**, 365–375.

Stuth, J.W., Grose, P.S. and Roath, L.R. (1987). Grazing dynamics of cattle stocked at heavy rates in a continuous and rotational grazed system. *Applied Animal Behaviour Science*, **19**, 1–9.

Tarrant, P.V. and Kenny, F.J. (1986). Methods for the study of cattle behavior during road transport. *Annales de Recherche Vetérinaire*, **17**, 100–101.

Tilbrook, A.J., Cameron, A.W.N. and Lindsay, D.R. (1987). The influence of ram mating preferences and social interaction between rams on the proportion of ewes mated at field joining. *Applied Animal Behaviour Science*, **18**, 173–184.

Van Liere, D.W., Kooijman, J. and Wiepkema, P.R. (1990). Dustbathing behaviour of laying hens as related to quality of dustbathing material. *Applied Animal Behaviour Science*, **26**, 127–141.

Van Rooijen, J. (1991). Feeding behaviour as an indirect measure of feed intake in laying hens. *Applied Animal Behaviour Science*, **30**, 105–115.

Walton, R.A. and Hosey, G.R. (1984). Observations on social interactions in captive Pere David's deer (*Elaphurus davidianus*). *Applied Animal Ethology*, **11**, 211–215.

Weiguo, L. and Phillips, C.J.C. (1991). The effects of supplementary light on the behaviour and performance of calves. *Applied Animal Behaviour Science*, **30**, 27–34.

Widowski, T.M. and Curtis, S.E. (1990). The influence of straw, cloth tassel, or both on the prepartum behaviour of sows. *Applied Animal Behaviour Science*, **27**, 53–71.

Workman, L. and Andrew, R.J. (1989). Simultaneous changes in behaviour and in lateralization during the development of male and female domestic chicks. *Animal Behaviour*, **38**, 596–605.

Zito, C.A., Wilson, L.L. and Graves, H.B. (1977). Some effects of social deprivation on behavioural development of lambs. *Applied Animal Ethology*, **3**, 367–377.

8

Companion Animals

H.M.R. Nott and J.W.S. Bradshaw

8.1 INTRODUCTION

For the majority of people in developed countries the closest contacts they have with any animal are with pets. Pets are kept for a variety of reasons, including pest control, security, routine or exercise. However, it is now clear that the majority are kept for pleasure rather than profit, and the primary motive for pet ownership is companionship (Endenberg *et al.*, 1990). Companion animals include species such as cats and dogs which have been associated with man for thousands of years, and also those that were previously domesticated for other reasons, such as rabbits and rats. In addition, there are also the more exotic species which have undergone little, if any, domestication, such as the larger psittacines and many aquatic species. This review will focus primarily on cats and dogs, since these are the most abundant companion animals, although reference will be made to other species where relevant studies have been conducted.

It is clear from archaeological and historical data that dogs, and later cats, were the earliest species reared as companions and that, additionally, they have evolved both physically and behaviourally as a result of man's selection. As a consequence, many of their natural behaviour patterns are a result of their association with man, and indeed it is hard to assign a natural habitat for these species other than as part of human society. Despite this close association with man, and the abundance of both cats and dogs throughout the world, there have been relatively few detailed studies of their ecology and behaviour. This trend is, however, changing, and more recently there has been an increased research emphasis placed on understanding the behaviour of companion animals both in relation to their ancestry and in their relationships with man (Turner and Bateson, 1986; Serpell, 1993).

Video Techniques in Animal Ecology and Behaviour. Edited by Stephen D. Wratten. Published in 1993 by Chapman & Hall, London. ISBN 0 412 46640 6

Studies on companion animal behaviour can be divided into five main areas of research: home ranges and territories, social behaviour, feeding behaviour, behavioural development, and their interactions with man. In this review we will consider each of these in turn and highlight the relative advantages and disadvantages of using video techniques as a research tool.

8.2 HOME RANGES AND TERRITORIES

There have been a large number and variety of studies on the behaviour of free-living cats, particularly feral and farm cats that do not rely solely on humans providing food. In different studies the density of cats varied from one animal per square kilometre to approximately 2000. The lowest cat densities tend to be found where the cats survive mainly on prey, whereas the highest densities occur where cats have food provided by humans, or scavenge at reliable refuse dumps. When animals are at high densities they tend to have smaller territories and home ranges, and so in general are easier to observe, for assessment both of their territory size and their social behaviour.

Studies on the home ranges and territories of feral and free-living dogs (which are owned but allowed to roam freely) have also shown that densities and home ranges change in relation to the availability of food. However, pack sizes tend to be larger where densities are lower (Nott, 1992), possibly to ensure that the group can defend those reliable food supplies that do exist (Macdonald and Carr, 1993). In general, researchers investigating the home ranges and territories of both cats and dogs have not used video techniques due to the large areas that have to be monitored. Instead they have relied on the use of radiotelemetry (see Amlaner and Macdonald (1980) for a review of methods) or the presence/absence of animals on different days or nights at specific sites by direct observation. In these latter cases direct observation has been used because of the wide area that needs to be monitored in detail, along with the advantage that more detailed observations of the social behaviour of any individuals present can also be made.

It is still possible to observe the behaviour of animals within a predefined area by making video records. For example, strategically placed cameras could monitor the movement of free-living cats or dogs along a pathway, or their presence or absence from a resting site. This technique has been used to observe the movements of free-living wild rats between two spatially distinct burrow groups (Nott, 1988). A monochrome time-lapse video recorder (Panasonic 6010) was used with a video camera (Hitachi Remote Eye) mounted on a tripod in a weatherproof box. A 30 W red bulb in an outside bulkhead light provided sufficient light to allow continuous recording without disturbing the animals. By using an auto-iris wide-angle lens on the camera no adjustments had to be made between day and night recording, as long as the camera was not positioned facing directly into the sun. The clarity of the image was sufficient to be able to distinguish between young and adult rats, but could only distinguish individuals if they had particularly obvious scars or characteristic postures. In general, dogs and cats are more physically distinctive than wild rats, and therefore individual recognition would probably be more feasible.

8.3 SOCIAL BEHAVIOUR

8.3.1 Simultaneous focal animal sampling

The behaviour of the domesticated carnivores is often sufficiently complex to warrant video recording even when only one animal is being observed, or when individuals from a group are observed one at a time, i.e. sequential focal animal sampling. Because companion animals are generally both accessible and approachable, at first sight it might seem that sequential sampling would be the most sensible way to collect information on social behaviour. However, it is not always possible to observe a group of dogs or cats for a sufficiently long period to obtain a representative sample of the behaviour patterns of each individual; for example, their owner may be unwilling to allow the observer sufficient uninterrupted access. In other situations, it may be necessary to obtain information about social behaviour that applies to a single point in time, such as during the development of a litter of puppies or kittens, or immediately following the introduction of a new individual into an established social group. In such cases video can be used to gather detailed information about several or even all members of the group simultaneously, depending on how detailed the recording of behaviour patterns needs to be, and therefore whether the camera can be trained on the whole group or only a subset.

The former is often the case when the social play of puppies is to be recorded, since prior to housebreaking they are often maintained within a comparatively small area for reasons of hygiene. Typically, the sleep/wake cycles of each member of the litter are roughly synchronized, with the effect that periods of inactivity, during which most of the puppies are asleep, alternate with periods of intense social play, involving all the litter-mates. During the latter, pencil and paper or event recorders are insufficient to record more than a small proportion of the detail available within the social interactions. Video recording in real time, with repeated slow-motion playback, can be used to produce a set of focal animal samples, one for each member of the litter. By this method, Hoskin (1991) could demonstrate that the social hierarchy within a litter of French bulldogs changed considerably from one session to another (Table 8.1), and although in this particular study the sessions were spaced at least 3 days apart, more detail of the changes in the hierarchy could have been obtained by recording every session of play that occurred. Certainly this analysis indicates that data from

Table 8.1 Changes in the hierarchy within a litter of five French bulldog puppies (three female, denoted F; two male, denoted M). Each hierarchy was deduced from video recordings of a single session of social play on each day; the rank orders were calculated from the ratio between the number of contacts given and received within each of the 10 possible dyads. In the first session the pups' coordination was still maturing, and it is possible that each hierarchy simply indicates that one pup was temporarily more active than the others. However, in the later sessions, when the hierarchies are more clear-cut, the relative rankings change from one session to the next. Further details of this study are given in Bradshaw and Nott (1993)

Age (days)	Hierarchy
28	Fp> others (circular)
33	Mb> others (circular)
41	Mb> others (circular)
44	Mb>Fp>Fb>Mm>Fm
47	Fp>Mb>Mm>Fb>Fm
54	Fp>Mm>Fb>Mb=Fm
63	Mb>Mm>Fm=Fp=Fb

different sessions cannot be pooled, which would be essential if simultaneous focal animal sampling was not possible.

Although social relationships are unlikely to change so rapidly in groups of adult dogs or cats, simultaneous recording of several individuals still has its advantages. First, the number of observation sessions can be reduced, which in turn reduces the probability that the presence of a camera or observer will disrupt or distort the behaviour being recorded. Second, it is usually much less disruptive to the owners or keepers of the animals if data recording can be confined to a small number of discrete sessions. Thirdly, for each set of simultaneous samples the environmental variables, such as the availability of food, warm or shaded resting places, and the proximity of conspecifics that are not part of the group, are the same. It is still uncertain how much of an effect each of these has on the social interactions of companion animals, but it is much easier to control for these by taking simultaneous samples than it is to arrange for conditions to be identical at the beginning of each of a set of a series of sequential samples.

By using this technique Wickens and Bradshaw (described in Bradshaw and Nott, 1993) have been able to demonstrate changes in the pattern of social interactions in a group of five adult French bulldogs, depending upon the context in which they occurred. The relative rankings of the four females, based upon the number of displacements given and received, were unaffected by context (15 different types of situations were tested, including food, toys, access to the owner's house, access to the owner, and other male and female dogs). However, the single male displaced all the females, including the alpha-, more often than they displaced him when the context was access to a familiar male dog of the same breed, or to a novel toy or object. His relationship with the alpha-female was reversed when they were competing over food. From the video recordings, it was also possible to discriminate between simple displacements, when one dog is pushed out of the way by, or defers to, another, and actual aggression. The latter was only common between the male, the alpha-female, and the beta-female (80% of aggressive interactions in 3/10 of the possible pairwise relationships).

In another breed, the Siberian husky, Wickens (unpublished) has observed much more complex social interactions, based on a repertoire of 42 behaviour patterns (Table 8.2), many of which are similar to those of the wolf, *Canis lupus*. Although these are large and active dogs, and it is rarely possible to hold all the members of the group within the field of view of the camera for any length of time, video recording is still a more efficient method of gathering data than direct observation. In particular, the ethogram which was used to define the patterns could be built up from the samples themselves, rather than a series of lengthy preliminary sessions in which no actual data could be gathered.

Table 8.2 Behaviour patterns used in describing social interactions within a group consisting of four male Siberian huskies, a male husky/wolf hybrid, and a female Eskimo sled dog, together with the number of occurrences of each recorded from approximately 10 hours of videotape

Active submission	8	Inhibited lick muzzle	9
Aggressive gape	512	Look away	43
Approach	224	Mount	26
Bark	31	Move away	329
Body slam	18	Muzzle bite	64
Ears back	629	Inhibited muzzle bite	66
Ears forward	377	Muzzle pin	12
Flank bite	133	Neck bite	529
Inhibited flank bite	63	Inhibited neck bite	251
Gape	766	Passive submission	73
Growl	11	Paw at	203
Hackles up	63	Play bow	69
Head/mouth wrestle	43	Stare	37
Head bite	111	Submissive gape	64
Inhibited head bite	357	Tail down	151
Hip slam	38	Tail up	215
Ignore	73	Threat	59
Leg bite	136	Throat bite	50
Inhibited leg bite	195	Inhibited throat bite	35
Lick head	15	Throat pin	29
Lick muzzle	34	Wag tail	117

Disadvantages

The technique of simultaneous focal animal sampling has both theoretical and practical disadvantages compared to the more conventional sequential sampling. The theoretical objection is that samples taken of two animals are not independent from one another, particularly if those two animals have actually interacted during the sampling period, and so statistical treatments may have to be modified, and in some cases the sampling time increased to compensate for the lack of independence. In practical terms, circumstances unknown to the experimenter and outside her control may affect all the behaviour seen at a single session, and therefore a whole set of focal animal samples rather than just one or two. Statistical analysis of session effects is therefore advisable before comparisons between animals can be made, and if large session-to-session variation is found, either the number of sessions may need to be increased or more attention paid to those factors that might be contributing to that variation.

8.3.2 Elimination of effects due to the presence of an observer

The social behaviour of domestic cats generally occurs at a much slower rate than that of dogs, and this may be one reason why, until comparatively recently, it was thought that behavioural interactions between cats were of little significance (Bradshaw, 1992). It is normally easier to record the social life of colonies of feral cats without recourse to video recording, but since cats are capable of interacting socially with people as well as animals, there is a possibility that the presence of an observer could affect even their interactions with other cats. This possibility can be eliminated by concealing the observer (who will be described as female in the following account). For example, cat colonies based around buildings will often appear to behave neutrally towards vehicles, even when these are occupied. Even if the observer is visible, dogs (Smith, 1983) or cats may not persist for long in trying to interact with her if their approaches provoke no reaction.

However, an advantage of remote video recording has emerged from a study of a colony of neutered cats interacting in a courtyard (Brown, 1993). On some days their behaviour was transcribed from video (with the disadvantage that not all of the area could be included in the field of view of the camera without losing vital detail), and on others the same observer sat on a high stool in the courtyard and noted the interactions directly. In the latter case, she did not reciprocate any interactions offered by the cats, and no records were taken for the first 5 minutes after she took up her position. The presence of the observer had a powerful effect on the frequency at which certain social behaviour patterns were expressed (Fig. 8.1). Defensive patterns (cuff and ears back) were more common when the observer was present, while the affiliative tail-up signal

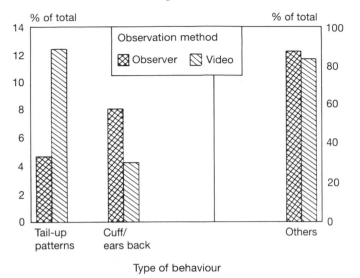

Figure 8.1 Behaviour patterns in a group of neutered cats, expressed as a proportion of the total number recorded, as observed directly ($n = 1377$) or transcribed from videotape ($n = 959$). Tail-up patterns were the combined totals of tail-up, tail-raised approach, tail-raised running, and tail-raised move away. The proportions of these four patterns were similar within the two methods of observation ($\chi^2 = 5.9$, 3 d.f., $p > 0.1$), but in total they were much less common when the observer was present ($\chi^2 = 43.4$, 1 d.f., $p < 0.0001$). The defensive patterns cuff and ears back could also be combined ($\chi^2 = 0.11$, 1 d.f., $p > 0.5$), and were more common when the observer was present ($\chi^2 = 9.75$, 1 d.f., $p < 0.005$). The remaining 28 common patterns were unaffected by the observation method ($\chi^2 = 36.0$, 27 d.f., $p > 0.1$).

was more common when she was absent. The reasons for this change are unclear at present, but there is no reason to suppose that similar effects do not also occur in other animals that are accustomed to the proximity of humans.

Hubrecht (1993) has monitored the behaviour of dogs in different kennelling conditions, including two shelter and two laboratory sites. He found that direct observation was possible at the animal shelters, probably because the dogs were habituated to the presence of the general public. However, at the laboratory sites he found it was necessary to use remote video recordings. Comparison of live observations versus video recordings of 10 dogs showed that single-housed dogs were significantly less active when a strange observer was in the room. These results demonstrate the need for preliminary analysis before relying on a single methodology in a variety of contexts.

8.4 FEEDING BEHAVIOUR

Bradshaw and Thorne (1992) have recently reviewed the published data on the feeding behaviour of cats and dogs. Domestic cats and dogs differ from their wild ancestors in that they have little choice about their type of food, as this is generally controlled by the owner, although there are obvious ways in which the pet's behaviour can influence the owner's buying decisions! The use of remote recording versus direct observations has a number of distinct advantages. First, the animal or animals can be observed for long time periods continuously, which would be possible but probably not practicable by direct observation. The second advantage is that more than one animal can be recorded simultaneously, and then replayed separately for individual analysis. This is particularly valuable in studies of social behaviour, when the frequency of events, e.g. meals, is high. Finally, probably the greatest benefit is that the animals are not influenced by the presence of an observer. Although not necessarily wary of humans, as would be the case for wild animals, domestic cats and dogs may feed differently in the presence of an observer. The behaviour of domestic dogs versus their wild counterparts is inextricably linked to the domestication process, which has resulted in the dog responding to humans as another sort of dog. This means that a dog may alter its behaviour while feeding, for example by waiting for verbal reassurance, guarding its food, or continually checking whether the observer is still present. Obviously it would be possible to use one-way glass or screens to minimize such effects, but the acute hearing and sense of smell in the dog can make this practically difficult.

Nott and Thorne (unpublished), studying the feeding behaviour of dogs, have found that the use of video techniques does not, however, automatically solve all problems. They were investigating the responses of dogs which had been reared on diets with differing degrees of variety, to novel food types. In general, dogs had been found to prefer an alternative diet after being maintained on a monadic diet for a number of weeks (Mugford 1977a,b). However, it has also been reported that dogs reared on distinctive diets differed in their responses to novel diets, with those reared on the widest variety of foods being the most willing to sample alternatives (Kuo, 1967; Ferrell, 1983). In addition to simply recording the amount of each novel food offered that the dogs consumed, they also recorded their latency to eat, whether they sniffed the food before starting to eat, and the amount of time spent eating. To avoid the possibility of contextual neophobia – in other words the dogs refusing the food because the environment had changed – it was necessary to habituate the dogs to the presence of the camera. In practice it was found easier to use a ceiling-mounted camera with a long lens rather than a camera on a tripod pointing over the kennel door, as this seemed to appear less obvious to the dogs and they habituated more rapidly. Despite this, at least three meals with the camera present were necessary to habituate the dogs and eliminate food guarding. In addition, the dogs were used to being fed individually penned, but still within their usual kennelling, which meant they were aware whether or not the other dogs in the building were feeding. To avoid this possibly biasing the results, each dog was tested for each novel food on a separate day. Thus, the use of video techniques enabled the data to be collected as required, but habituation was necessary, even if not to the extent of habituation to a human observer.

8.4.1 Continuous analysis of feeding patterns

Video techniques have also been utilized in the analysis of the ad-libitum feeding patterns of various species of companion animal. These studies have highlighted fundamental differences between species and, more interestingly, between different breeds of the same species. Cats have been found to take many small meals throughout the day and night (Fig. 8.2), although the exact number taken has varied from study to study, depending on the methodologies used (Table 8.3).

The incidence of obesity in dogs is much higher than in cats, and it would be expected that they could not regulate their food intake as closely as the cat. However, studies at the Waltham Centre (Mugford and Thorne, 1980) on beagles, basenjis and poodles showed that they too adopted a nibbling strategy – many frequent meals – when not constrained to feeding at specific times (Fig. 8.3). The beagles tended to feed on many small meals throughout the day and night, whereas the poodles and basenjis both adopted a diurnal pattern of feeding, although there was considerable individual variability. Despite this pattern of discrete meals, many of the dogs did not regulate their food intake according to their energy requirements, with the result that a proportion of the animals gained weight continuously until removed from the trial, or gained until they reached a new, higher, plateau.

The majority of domestic cats regulate their food intake quite closely and will usually maintain their body weight even when palatable food is freely available. Despite this, analysis of the feeding patterns of cats has not demonstrated a clear mechanism of food intake regulation, since no significant correlations between the length of a meal and the preceding or succeeding gap between meals have been consistently found. One possible reason for this lack of consistency includes the environmental conditions in which the animals were housed, and in particular whether they were housed singly or in a group. Slater (1981), studying the feeding of zebra finches, has observed significant differences between birds housed singly and those in a group. Group-housed birds were less likely to show significant prandial correlations than singly housed birds, and Slater suggested that this was a function of the social status of the birds, with dominance disrupting their normal feeding behaviour. Unfortunately he had no behavioural data to confirm this, but recent studies on wild rats (Nott, 1988) have supported this explanation. Another explanation for the lack of consistency between the cat studies are the methodologies employed. Some workers had used video techniques, whereas others had used photocells or load cells connected to

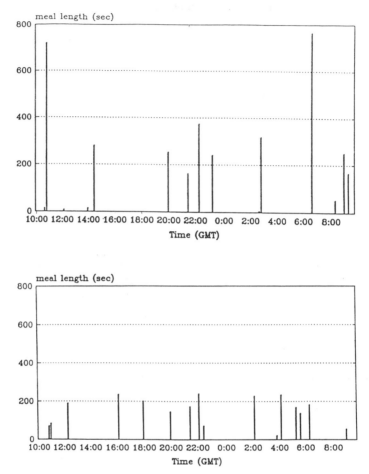

Figure 8.2 Feeding patterns of two individually housed adult male cats shown for representative 24-hour periods. Meal length was defined as time spent feeding at the food bowl, with at least a 3-second interval of head-up defining a gap between two meals. Both cats were maintained on a complete dry diet, Whiskas Crave (Kal Kan, USA).

an event recorder or computer. Replaying from video tapes and data from photocells measured meal size in terms of duration, whereas load cells measure meal size in terms of food eaten. Physiologically it is the amount eaten that is going to be most important in any regulatory mechanism for ingestion. Therefore, the use of load cells is probably most appropriate, although video records or a transponder system may be a necessary adjunct when animals are group-housed. Care should, however, be taken in the analysis of automatically recorded data since, unlike with video records, you cannot be sure that the animal has not dropped food which it then eats later,

Table 8.3 Summary of studies which have examined the feeding patterns of cats with ad-libitum access to food. Correlations are for each meal with either the preceding or succeeding gap between meals

Reference	Method	Conditions	Diet	Mean no. of meals	Significant correlation Pre	Post
Kanarek, 1975	photocells	metabolism cages	ground Purina cat chow	8–12	yes	no
Mugford, 1977a, b	video	group	Kitekat Munchies[a]	13.5		
Mugford and Thorne, 1980	video	group	Brekkies[a]	7.3–15.6	yes	no
Kaufman et al., 1980	photocell	uncaged, single	?	10	no	no
Kane et al., 1981	load cell	caged	Crave[b]	15.7 ± 1.8	yes	no
			Seaside supper[b]	16.6 ± 1.8	yes	no
			purified:			
			casein-based	16.8 ± 2.3	no	no
			amino acid-based	17.4 ± 3.5	no	no
Kane et al., 1987	load cell	caged	purified:			
			high fat	10.4 ± 1.1		
			low fat	11.4 ± 1.2		
Nott (unpubl)	video	uncaged, single	Crave[b]	17.6 ± 4.4	no	no

[a] Pedigree Petfoods, UK
[b] Kalkan Foods, Los Angeles, USA

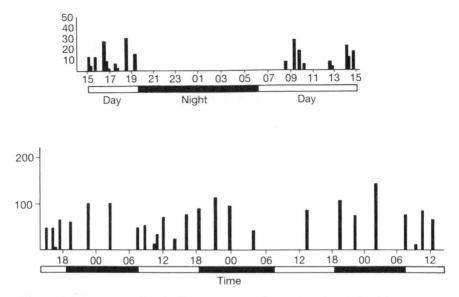

Figure 8.3 Representative feeding patterns of two breeds of dog, basenjis and beagles. Both dogs were maintained on the same complete dry diet. Data for the basenjis is a single 24-hour period, and that for the beagles is 72 hours. A meal is defined as a bout of feeding followed by a minimum period of 5 minutes non-feeding (from Mugford and Thorne, 1980).

or is not sniffing food without actually feeding. Permanent video records allow rational decisions to be made as to when a meal started or finished, and therefore provide a good basis for validating any automatic systems. For example, Munday and Earle (1991) have used video records to verify the accuracy of a system designed to measure the food intake of a queen separately from her kittens during weaning. They were able to establish when the kittens were able to learn the system and climb sufficiently to reach the queen's food.

8.5 BEHAVIOURAL DEVELOPMENT

One area in which video techniques have been quite extensively used is in the study of development. Continuous time-lapse video records allow the first occurrence of an event in behaviour to be observed, resulting in more precise timings than simply the nearest day on which observations were carried out and a behaviour was first seen. Such accuracy can be particularly important if comparisons in the rate of development are to be made between individuals in a litter, or between litters.

Thorne *et al.* (1993) have continuously monitored the behaviour of a queen from 2 weeks before parturition until the kittens were 6 weeks old. They were also able to monitor the changing time budgets as the animals developed. They were able to observe when the kittens first started to clean themselves when 24 days old, and the gradual reduction in the frequency with which the queen cleaned them. In addition, the queen was observed moving the kittens out of the kittening box at a regular time each day, an activity that would have gone unobserved in any focal study. In this preliminary study, with only one litter, they also found that many of the first occurrences of a behaviour were earlier than in previously published accounts, confirming the benefits of continuous versus focal observation.

Grant (1987) continuously monitored the behaviour of a beagle bitch and her litter from whelping for a further 20 days. A low-light infra-red-sensitive camera (Sanyo) was used to allow filming at night. Changes in maternal behaviour from whelping to the start of weaning were recorded, as well as the way in which the feeding and cleaning activity varied between daytime and night-time. One of the most interesting findings of this study was the way the bitch changed her behaviour in response to the presence of people during the day. During working hours she took up a sitting posture while her pups were feeding whereas outside working hours she was generally more relaxed and would lie down to feed and clean her pups. In a subsequent study, Grant and Rainbird (1987) monitored the suckling behaviour of puppies in three litters of labradors. Again, they found that the bitches sat up to feed their puppies in the daytime, particularly when there was a lot of activity in the building. In the night the bitches lay down to feed and, by recording on which side the bitches lay, they found that the length of time the bitches spent feeding was divided fairly evenly between time spent lying on the left and right hand side (Fig. 8.4).

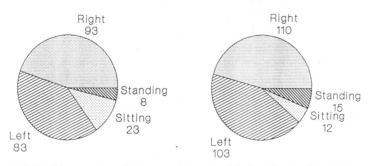

Figure 8.4 Positions adopted by two labrador bitches for feeding their puppies. Data are the total numbers of feeds at each position over the entire duration of the investigation (25 days). Right and Left refer to the side the bitch was lying on when lying to feed her puppies (from Grant and Rainbird, 1987).

8.6 INTERACTIONS WITH MAN

The behaviour of companion animals that are interacting with their owners is likely to be just as complex as that of companion animals interacting with each other, but to date video recording has not been used extensively in this type of study. Both Smith (1983), studying dogs within their families, and Mertens and Turner (1988), studying cats in staged encounters with unfamiliar people, used video recording as a backup to direct observation, but make no mention of any special advantages of the recordings. In staged encounters between owners, or strangers, and groups of six dogs, Brown and Bradshaw (unpublished observations) analysed the interactions between a focal dog that was being petted, and the other five. Video recording was facilitated by instructing the human to conduct his or her interaction with the dogs at a fixed point, and data on the proximity of each dog to the human, and the attempts made by each dog to gain the attention of the person, could both be recorded from the videotape.

Eckerlin and coworkers (1989) have analysed the relationship between children and their pet dogs using a structured paradigm. They were particularly interested in the acoustic cues used in communication and their effect on the dogs' behaviour. Different tape recordings of the voices of a familiar child or an unknown child were played in the presence of the dog through loudspeakers concealed in two dummies. The behaviour of the dogs was recorded for 5 minutes using video film, and the tapes were then analysed to provide data which quantified the different behaviours of the dog according to the characteristics of the voice that was presented. Obviously in this study it was important that a visible observer was not present, and video techniques have the additional advantage of allowing detailed analysis of such rapidly changing behaviours. This type of investigation, using ethological techniques rather than the more traditional psychological techniques, represents an increasing trend in studies that aim to investigate in more detail the reasons and mechanisms behind human–animal relationships (Hart, 1989).

Unstructured, spontaneous interactions between companion animals and humans may be less easy to record. Most occur in unpredictable locations indoors, and are therefore unsuitable for remote recording. The presence of a camera operator is likely to affect the behaviour of the human partner, and possibly that of the animal, to the same extent as an observer using an electronic event recorder, and so the speed with which data gathered by the latter method can be analysed is likely to have been the deciding factor when studies of this type have been carried out (e.g. Mertens, 1991).

8.7 GENERAL CONSIDERATIONS

There is no real doubt, from the variety of studies described, that the applications of video techniques in the study of companion animal behaviour are vast. The main advantages are the reduction in disturbance versus direct observation, and the ability to record continuous data for long time periods that would be possible, but not really practicable, by direct methods. This latter advantage allows infrequent events to be monitored, allows detection of the first occurrence of events during development, and allows simultaneous focal animals sampling to be conducted from social interactions in groups. In addition, permanent video records allow the replay of events where the sequence of behaviours, or behaviours themselves, is unclear. The other advantages of these methods are that optimal sampling methods can be selected, and that they can be used as a check for automatic recording systems.

One of the main disadvantages of using video is the limited field of view of the camera. Even if wide-angle lenses are used the two-dimensional image can often make it difficult to determine if two animals are actually in contact, and whether or not an animal is actually feeding or simply sniffing the food. The other main disadvantage, as already mentioned, is the tendency to rely on one period of behaviour that has been analysed in detail. This can be minimized by recording other sessions and analysing more briefly to confirm the validity of the detailed session; however, this may not be practical when spontaneous encounters are being examined, and can result in many hours spent playing back videos! The only other factor that has limited the use of video techniques is the technical considerations. The lack of mains electricity or a secure place in which equipment can be housed, although not necessarily insurmountable, are factors that need to be considered before opting to use video techniques versus traditional observational methods.

Acknowledgements

JWSB would like to thank the Waltham Centre for Pet Nutrition and the Science and Engineering Research Council for financial support. We would like to thank Helen Brown, Sarah Brown, Chris Thorne and Steve Wickens for permission to quote unpublished results, and Pam Wellstead, Sue Hull and Sandy Hawkins for access to dogs.

REFERENCES

Amlaner, C.J. and Macdonald, D.W. (1980). *A Handbook of Biotelemetry and Radiotracking*, Pergamon Press, Oxford.
Bradshaw, J.W.S. (1992). *The Behaviour of the Domestic Cat*, CAB International, Wallingford, UK.
Bradshaw, J.W.S. and Nott, H.M.R. (1993). Social and communication behaviour of companion dogs, in *The Domestic Dog: its Evolution, Behaviour and Interactions with Man* (ed. J.A. Serpell), Cambridge University Press, Cambridge (in press).

160 *Companion animals*

Bradshaw, J.W.S. and Thorne, C.J. (1992). Feeding behaviour, in *The Waltham Book of Dog and Cat Behaviour* (ed. C.J. Thorne), Pergamon Press, Oxford, pp. 115–130.

Brown, S. (1993). *The Social Behaviour of Neutered Domestic Cats* (Felis catus). PhD Thesis, University of Southampton.

Eckerlin, A., Filiatre, J.C., Millot, J.L. and Montagner, H. (1989). An ethological approach to the acoustic cues in the relational systems between the child and his pet dog. Paper presented at *Vth International Conference on the Relationship Between Humans and Animals*, Monaco 1989.

Endenberg, N., Hart, H. and de Vries, H.W. (1990). Differences between owners and nonowners of companion animals. *Anthrozoös*, **4**, 120–126.

Ferrell, F. (1983). Early flavor experience and postweaning food preference in puppies. Presentation for the *PFI Technical Symposium at the Fourth Annual Convention of the Pet Food Institute*, Washington DC, September 23 1983.

Grant, T.R. (1987). A behavioural study of a beagle bitch and her litter during the first three weeks of lactation. *Journal of Small Animal Practice*, **28**, 992–1003.

Grant, T.R. and Rainbird, A.L. (1987). A behavioural study of three bitches and their puppies from onset of weaning to six weeks of age, in *Nutrition, Malnutrition and Dietetics in the Dog and Cat, Proceedings of an International Symposium*, Hannover, pp. 134–135.

Hart, L. (1989). Ethology applied to human–animal interactions: research techniques from the field of animal behaviour. Paper presented at *Vth International Conference on the Relationship Between Humans and Animals*, Monaco 1989.

Hoskin, C.N. (1991). *Development of the Dominance Hierarchy Amongst a Litter of French Bulldog Pups*. BSc thesis, University of Southampton.

Hubrecht, R. (1993). The welfare of dogs in human care, in *The Domestic Dog: its Evolution, Behaviour and Interactions with Man* (ed. J.A. Serpell), Cambridge University Press, Cambridge (in press).

Kanarek, R.B. (1975). Availability and caloric density of the diets as determinants of meal patterns in cats. *Physiology and Behaviour*, **15**, 611–618.

Kane, E., Leung, P.M.B., Rogers, Q.R. and Morris, J.G. (1987). Diurnal feeding and drinking patterns of adult cats as affected by changes in the level of fat in the diet. *Appetite*, **9**, 89–98.

Kane, E., Rogers, Q.R., Morris, J.G. and Leung, P.M.B. (1981). Feeding behavior of the cat fed laboratory and commercial diets. *Nutrition Research*, **1**, 499–507.

Karsh, E.B. and Turner, D.C. (1988). The human–cat relationship, in *The Domestic Cat: the Biology of its Behaviour* (eds D.C. Turner and P. Bateson), Cambridge University Press, Cambridge, pp. 159–178.

Kaufman, L.W., Collier, G., Hill, W.L. and Collins, K. (1980). Meal cost and meal patterns in an uncaged domestic cat. *Physiology and Behaviour*, **25**, 135–137.

Kuo, Z.Y. (1967). *The Dynamics of Behaviour Development: An Epigenetic View*, Random House, New York.

Macdonald, D.M. and Carr, G. (1993). Free-living dogs in the Abruzzi: the behaviour and ecology of contrasting populations, in *The Domestic Dog: its Evolution, Behaviour and Interactions with People* (ed. J. Serpell), Cambridge University Press, Cambridge (in press).

Mertens, C. (1991). Human–cat interactions in the home setting. *Anthrozoös* **4**, 214–231.

Mertens, C. and Turner, D.C. (1988). Experimental analysis of human–cat interactions during first encounters. *Anthrozoös*, **2**, 83–97.

Mugford, R.A. (1977a). Comparative and developmental studies of feeding behaviour in dogs and cats. *British Veterinary Journal*, **133**, 98.

Mugford, R.A. (1977b). External influences on the feeding of carnivores, in *The Chemical Senses and Nutrition* (eds M.R. Kare and O. Maller), Academic Press, New York, pp. 25–50.

Mugford, R.A. and Thorne, C.J. (1980). Comparative studies of meal patterns in pet and laboratory housed dogs and cats, in *Nutrition of the Dog and Cat* (ed. R.S. Anderson), Pergamon Press, Oxford, pp. 3–14.

Munday, H.S. and Earle, K.E. (1991). Energy requirements of the queen during lactation and kittens from birth to 12 weeks. *Journal of Nutrition*, **121** (Suppl), 43–44.

Nott, H.M.R. (1988). *Dominance and Feeding Behaviour in the Brown Rat.* PhD thesis, University of Reading.

Nott, H.M.R. (1992). Social behaviour of the dog, in *The Waltham Book of Cat and Dog Behaviour* (ed. C.J. Thorne), Pergamon Press, Oxford, pp. 97–114.

Serpell, J.A. (ed.) (1993). *The Domestic Dog: its Evolution, Behaviour and Interactions with Man*, Cambridge University Press, Cambridge.

Slater, P.J.B. (1981). Individual differences in animal behaviour, in *Perspectives in Ethology* Volume 4; *Advantages of Diversity* (eds P.J.B. Slater and P.H. Klopfer), Plenum Press, New York, pp. 35–49.

Smith, S.L. (1983). Interactions between pet dog and family members: an ethological study, in *New Perspectives on Our Lives with Companion Animals* (eds A.H. Katcher and A.M. Beck), University of Pennsylvania Press, Philadelphia, pp. 29–36.

Thorne, C.J., Mars, L.A. and Markwell, P.J. (1993). A behavioural study of the queen and her kittens. *Animal Technology*, **44**(1), 11–17.

Turner, D.C. and Bateson, P. (eds) (1986). *The Domestic Cat: the Biology of its Behaviour.* Cambridge University Press, Cambridge.

9

Video and microorganisms

K. Sugino

9.1 NEED FOR VIDEO IN MICROBIOLOGY

The greatest difference between motion recording systems, such as cinematography and video recording, and conventional picture recording using microscopic photography, is the acquisition of a time dimension. Temporal data is, of course, very important in behavioural research. Nowadays, behavioural investigations, especially into organellar motility, are needed also in the taxonomic or morphological sciences for microorganisms (Inouye and Hori, 1991). For taxonomic studies on the evolution of protists, for example, morphological studies are not sufficient: when the research objects become more primitive, motile specificity should also be considered. Even for physiological research into protozoa there is insufficient ethological observation (Ricci, 1990).

Microbiology has severe handicaps though, in particular the size of research objects and the speed of their motion. Because of the small size of microbes, microbiological research on their behaviour is in most cases done under the microscope. The motion analysis of a single microorganism, beginning with so-called 'single cell tracking', has so far proved technically very difficult. The single cell observation of phototactic behaviour, for example, was almost impossible without a video system (Iwatsuki and Naitoh, 1981, 1983), because the behaviour itself is largely influenced by the strong illumination required for the film recording. It is not exaggerating to say that the combination of new video technology armed with highly sensitive high-speed cameras and computerized analysis techniques has brought in a new age of microbiology (Allen and Allen, 1981).

Video Techniques in Animal Ecology and Behaviour. Edited by Stephen D. Wratten. Published in 1993 by Chapman & Hall, London. ISBN 0 412 46640 6

This chapter will explain how video systems are used to observe the behaviour of microorganisms, and what kind of physiological information the behavioural video recording of such tiny creatures provides.

9.2 ADVANTAGES AND DISADVANTAGES OF VIDEO RECORDING

9.2.1 Advantages

The difference between the principles of image recording by video and by cinematography directly influences the style of microbiological research. The rewind and replay functions of the video recording system permit investigators to check their results during or immediately after the experiments, and then, if necessary, the experiments can be repeated with the same objects. On the other hand, experimental results recorded on cine film can be viewed usually only after several days. Because the video image is recorded electromagnetically, the recorded image can easily be converted with minimal deterioration into digital data for computerized analysis, by which the recorded images can be compared and evaluated quantitatively. Even excluding the easy transformation and transport of the digitized data for further image analysis, the merit of the video system is still very great. Because of the reusability of the recording medium – magnetic video tape – there is no need to wait for a 'shutter chance', upon which one has to concentrate. Especially with high-speed recording, one would have to use many cine films, which usually have to be highly sensitive for such purposes and hence very expensive. The merit of the video tape is evident when the recording is more orientated to observation than to experimental documentation. If the picture quality can be reduced slightly, the 'long play' mode of a VHS recorder, for example, permits continuous recording for 12 hours using a 240 min tape, and important scenes can be edited afterwards on to another video tape. Additionally, a video tape is usually set on the recorder, not on the camera, and can therefore be mounted and changed more easily and quickly than cine film. Because cine films must usually be mounted in a dark room, the experiment cannot continue when the film ends. Because of its recording mechanism, the film must be set into the camera and hence the camera, which is very close to or often mechanically connected to the microscope, must be demounted from the experimental setup in order to change the film each time.

One of the greatest progressions in motion recording technique is the high sensitivity of the video camera, the charge-coupled device (CCD). Even video cameras for domestic use, with a frame rate of 30 Hz, are as sensitive as human eyes. This means one can record whatever one can see, even under the microscope, without special illumination. Using a more sensitive camera, an additional video enhancer, strobe illumination or a combination of these, one can still take dark views such as fluorescence, at high speed.

The electronic shutter release of the video camera offers two further merits. One is the disappearance of mechanical noise, which is fatal for microscopic work. This specificity, together with the light weight of the video camera separated from the recording device, enables the camera to be mounted directly on to the microscope, which frees the investigator from refocusing each time. Another merit is the possibility of synchronous recording and/or frame combination of two video signals from two cameras. The multiple signal input is not yet easily available, but some cameras for industrial use have synchronous trigger inputs and outputs. This is very helpful in many kinds of experiment to synchronously monitor something other than the object organism, such as temperature, membrane potential, pH and so on. Superimposers to combine multiple frames have become popular.

9.2.2 Disadvantages

The weakest point in video recording is the low resolution of the image compared to cinematography. The vertical picture resolution is restricted, first of all, by the number of scanning lines. This will be overcome by the 'high vision' technique in the near future.

A more basic problem is the speed and capacity of the recording medium for image information. Because of the limited recording speed of magnetic tape, most high-speed video recorders use the interlacing picture fields as two independent frames. Sometimes a single picture frame is divided into several zones in order to follow the recording speed. This, of course, causes a strong reduction in spatial resolution in exchange for a high temporal resolution. Instead of magnetic tape, frame memories (RAMs) can be used for image storage, at very high frame rates, with sufficient spatial resolution, but in this case recording is possible only for a very short time. On the practical use of video microscopy, attention should be paid to the electrical noise of the camera device, especially if strobe illumination is used. Very strong noise may sometimes cause an error in vertical or horizontal synchronization.

9.3 VIDEO APPLICATIONS IN BEHAVIOUR ANALYSIS OF FAST-SWIMMING MICROORGANISMS

9.3.1 Observation of swimming microorganisms

The behaviour of fairly large, fast-swimming protozoa, such as ciliates, can easily be observed and analysed using a video system. The replay function is very useful for measuring and evaluating the behaviour of large numbers of microorganisms. When the data on large numbers of object microorganisms are obtained from a single trial, the level of experimental error will be greatly reduced. When only one measurement is possible for one trial, the same experiment must be repeated to satiety in order to accumulate the required amount of data and such repetition causes errors, because it is very difficult to keep the same experimental conditions and to repeat the experiment in exactly the same manner.

The first example is a setup to record the backward swimming (avoiding reaction) of ciliate *Paramecium caudatum* cells in a high K^+ concentration medium under controlled temperature conditions (Fig. 9.1). Without a visual recording possibility one would have to repeat the same experiment many times, since the duration of backward swimming can be measured for only one cell in a single experimental trial.

Because extremely high magnification is not necessary and a wide area should be observed, a video camera is set directly above the experiment vessel (Fig. 9.2), simply using a zoom lens or a bellows with a C-mounting connector.

The most important point is to take the video picture under dark-field conditions. For this reason the light illuminates the object organisms aslant to the optical axis of the video camera, and the inside of the water bath is painted with non-reflecting black. Although the upper illumination causes a dark-field effect, light from behind the vessel is recommended to prevent reflection on the water surface. If backlighting is not possible, a polarized filter for the video camera helps, but the image intensity will be halved.

If a conventional video camera is used it should be focused manually, because of the tendency of automatic focusing to focus on the water surface.

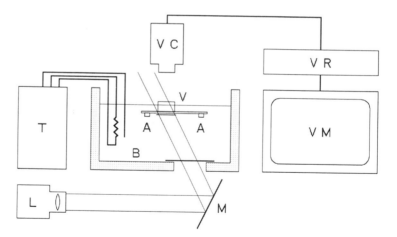

Figure 9.1 A simple setup of recording the swimming behaviour of *P. caudatum*. A: arms to hold the vessel; B: water bath to keep the temperature constant; L: light source; M: mirror; T: temperature control unit; V: experimental vessel (for detail, see Fig. 9.2); VC: video camera; VM: video monitor; VR: video recorder (after Y. Shingu, Thesis, 1989).

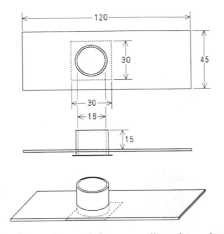

Figure 9.2 The experimental vessel for recording the swimming behaviour of *P. caudatum*. The bottom of the cylindrical experiment chamber is made from a cover glass in order to guide the illumination from the bottom. The backward swimming is caused by the high K^+ concentration experimental medium (20 mM KCl, 1 mM $CaCl_2$, 1 mM Tris-HCl, pH 7.2) in the chamber and is seen immediately after pouring the cells together with 20 µl of adaptation medium (4 mM KCl, 1 mM $CaCl_2$, 1 mM Tris-HCl, pH 7.2) into the chamber filled with the experimental medium. To observe only the backward-swimming organisms, the circular shape of the chamber is helpful, since forward-swimming *Paramecium* tends to accumulate at the edge. (Units: mm; after Y. Shingu, Thesis, 1989.)

In the recorded image, cells are recognized as moving points. In this example the duration of backward swimming and its velocity are measured.

9.3.2 Recording of tactic behaviour

One of the typical areas of behavioural research on microorganisms is into tactic behaviour (Van Houten *et al.*, 1975, 1981). Although a number of studies have been carried out, most were concerned with the description of tactic responses – accumulation or dispersal – and did not touch upon its mechanism. For precise analysis of the tactic behaviour of microorganisms, a long period of continuous recording with adequate time resolution is required. This section describes the video setup and analysis sequence for the phototactic behaviour of the ciliate *Paramecium bursaria*.

The recording of phototactic behaviour is sometimes faced with a dilemma, because the recording illumination also acts as a light stimulus. All the experiments must proceed in principle under dark conditions. Figure 9.3 shows the setup for the phototaxis experiment and its connection diagram. This is for a wide observation area and hence uses low magnification. The experiment stage, light source and other optics around it are set in a dark

A

B

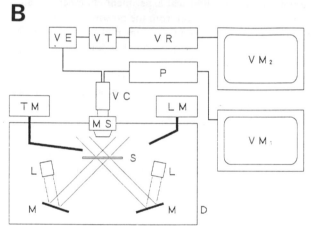

Figure 9.3 The setup for the phototaxis experiment with *P. bursaria* (A) and its connection diagram (B). D: dark box; L: light sources divided and led from the same halogen lamp (200 W) through glass fibres; LM: illuminance meter (Lux meter); M: mirrors, on which a heat ray filter is mounted; MS: low magnification microscope; P: power supply and video camera control unit; S: the light exposing stage (for detail, see Fig. 9.4) on which the experiment vessel is held; TM: thermometer; VC: CCD video camera; VE: video enhancer; VM_1: video monitor for the original picture; VM_2: video monitor for the recorded image; VR: video recorder; VT; video timer (after M. Kusaka, thesis, 1992).

box. Even though the video camera (CCD type) is very sensitive, a video enhancer or an image intensifier helps to provide high contrast, which makes it easy to binarize the image of the dark-field chamber with many cells. In this example the light intensity ranges from 250 to 38 000 Lux. If possible, a dual-monitor system is helpful for watching the actual chamber

(Fig. 9.3, VM$_1$) in addition to that for controlling the image intensity for optimized video recording (Fig. 9.3, VM$_2$).

When the behaviour of microorganisms as a reaction to a certain stimulus is recorded, the radiant heat generated, together with light illumination for

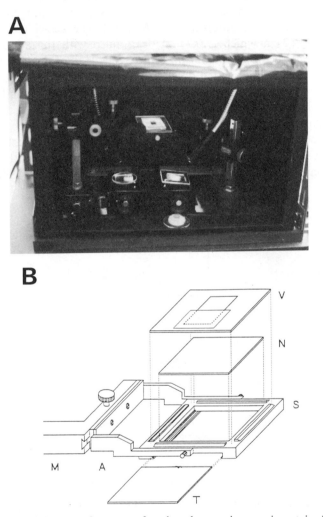

Figure 9.4 The light exposing stage for the phototaxis experiment in *P. bursaria* (A) and its schematic diagram (B). A: arm to hold the light exposing stage; M: manipulator; N: neutral-density filter to change the light intensity without changing the colour temperature; S: the light exposing stage, on which from the bottom a heat ray filter, neutral-density filter and the experimental vessel (for detail, see Fig. 9.5) are mounted. This stage is capable of rotation to keep the stage horizontal, in order to avoid the gravitactic influence on the swimming behaviour of the cells. T: heat ray filter to keep the chamber from temperature change (after M. Kusaka, thesis, 1992).

recording, should always be taken into consideration. In this setup two heat ray filters are used: one is mounted on the mirror (Fig. 9.3, M) and the other is set under the experimental vessel (Fig. 9.4, T).

This experiment is designed to investigate the relationship between photokinesis and photophobic response, i.e. photoaccumulation or dispersal, based on the chemotaxis experiment by Nakazato and Naitoh (1980, 1981, 1982) with some modifications (Fig. 9.5). The vessel has a chamber measuring 1×2 cm which is 0.7 mm deep, filled with the adaptation medium, in which the object organisms swim freely. Half of the chamber, i.e. an area of 1×1 cm, is illuminated differently from the other half, giving a light boundary in the middle. Photokinesis under light and dark conditions, and the photophobic response, are recorded synchronously in both areas and at the boundary of the two, respectively.

If the recorded area is not illuminated equally, as in this case, the image of the cells in the dark area sometimes disappears, and/or that in the light area shows a halo. In order to avoid this, the same neutral-density filter as that

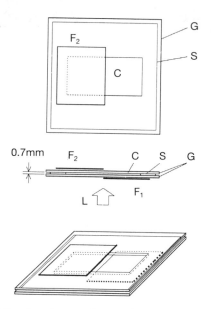

Figure 9.5 The vessel for the phototaxis experiment in *P. bursaria*. The swimming area in the chamber for the cells is 1×2 cm. Half of the area is darkened by a neutral-density filter below the vessel. Swimming in the perpendicular direction is restricted by two glass plates sandwiching 0.7 mm thick silicon rubber, leaving the swimming space free. C: swimming space for the cells; F_1: neutral-density filter to darken half of the chamber; F_2: neutral-density filter for image intensity compensation for recording; G: glass plates to seal the chamber; L: light for illumination and at the same time for the photostimulus; S: silicon rubber spacer, in the middle of which the two-dimensional swimming area is kept (after M. Kusaka, thesis, 1982).

for darkening is inserted as a dummy filter above the light area of the chamber to equalize the resulting image intensities in both areas without changing the light conditions to the cells.

When the light source for recording is used at the same time as the experimental stimulus, its intensity should not be changed by changing the voltage but by neutral-density filters to keep the temperature of the stimulus unchanged (Fig. 9.5, N).

9.3.3 Analysis of tactic behaviour

The kinetic behaviour of ciliate microorganisms is usually compared with Brownian movement or random walk motion, and can be evaluated by two parameters: the velocity of forward swimming and the frequency of the spontaneous avoiding reaction, which changes the swimming direction.

The photophobic response at a boundary is expressed as the 'passing ratio', i.e. the ratio of the number of cells passing over the boundary and not returning to that of the cells reaching the boundary.

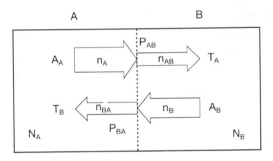

Figure 9.6 A basic model of the uneven distribution of microorganisms in a chamber with two different environments. N_A, N_B: counts of the cells in each of areas A and B. The total cell count N in the chamber is hence $N = N_A + N_B$; n_A, n_B: the rate of arrival of cells to the environmental boundary from each area, A and B in dimensions of $[T^{-1}]$; n_{AB}, n_{BA}: the rate of passing of the cells over the boundary from area A to B and vice versa $[T^{-1}]$; A_A, A_B: the 'arrival ratio' of the cells to the boundary in each area, A and B in $[T^{-1}]$ defined as

$$A_{A(B)} = \frac{n_{A(B)}}{N_{A(B)}};$$

P_{AB}, P_{BA}: the dimensionless passing ratio of cells over the boundary from each area, A to B, and vice versa, and defined as

$$P_{AB(BA)} = \frac{n_{AB(BA)}}{n_{A(B)}};$$

T_A, T_B: the 'transfer ratio', i.e. the possibility of crossing over the boundary in a unit time range, defined as $T_{A(B)} = P_{AB(BA)} A_{A(B)}$.

The cell distribution in the chamber (Fig. 9.6) is generally described as

$$\begin{cases} N_A(t)=(N_A(0)-T_B\tau N)e^{-t/\tau}+T_B\tau N \\ N_B(t)=(N_B(0)-T_A\tau N)e^{-t/\tau}+T_A\tau N, \end{cases}$$

where τ is the time constant. In this kind of experiment, in order to compare the cell densities in the two areas as the result of tactic behaviour, the number of cells in a unit area on each side of the chamber is usually counted after reaching stabilization. The 'attracting ratio' or the 'attraction index' is estimated from the cell counts in the two areas in the stationary state. With additional data from the cell count during transition, the time constant of cell accumulation and the 'transfer ratio' as the multiplication of arrival ratio and passing ratio can be estimated (see Appendix).

Although most parameters determining accumulation behaviour are estimated from such snapshot type data, in order to investigate its detail, e.g. to separate the transfer ratio into the arrival ratio and the passing ratio, the behaviour of each cell must be observed at the boundary of its environment. Above all, this general equation contains no time lag for adaptation. In other words, the cell is regarded as being capable of changing its kinetic behaviour, i.e. swimming velocity and turning frequency, without any delay after entering a new environment. For the precise analysis of such experiments, cell tracking should be carried out in parallel with the measurement of distribution.

One of the easiest ways of cell tracking using video recording data is to superimpose the video frames for a certain time range, for example 10 seconds, using a computer through a video converter or a video superimposer for professional use. If no computer is available, the superimposition of video pictures on a high-contrast 35 mm film with a conventional camera can be used instead (Fig. 9.7). The evaluation of the cell tracking data, e.g. to obtain the swimming path length, is usually by automated or manual measurement of the path. If no automated system is available, if the computer has a superimposition card, even a simple program showing just a cursor on the monitor displaying the video frame as background, and writing the position data (coordinate data) to a file on request, is useful enough. Already some sophisticated computer hardware and software have been developed for this purpose (Häder and Lebert, 1985, for *Euglena* phototaxis; Machemer *et al.*, 1991, for *Paramecium* gravitaxis; Eggersdorfer and Häder, 1991, for *Prorocentrum* phototaxis and gravitaxis). These programs calculate direction as well as velocity to show the swimming tendency in a form of polarogram or radar chart. Recently hardware and software systems for real-time multicell tracking have been developed (Katz and Overstreet, 1981; Dusenbery, 1985; Donovan *et al.*, 1986; Hasegawa *et al.*, 1988) and are commercially available, e.g. Cell Soft 3000 by Clio Research Co. for the IBM PC.

Figure 9.7 Cell tracking using superimposition of video frames by a film camera on a high-contrast 35 mm film. Eleven frames with intervals of 1 s were superimposed, i.e. 10 s from the beginning to the end. This gives the same effect as in the stroboscopic picture (Naitoh and Kaneko, 1972). The mean swimming velocity of 0.442 mm/s was measured from the data under light conditions of 2750 Lux. The original data and the superimposed picture were taken by M. Kusaka.

If the behaviour of cells at the boundary can be recorded, the arrival ratio and the 'responsiveness' are very easy to determine. The former depends on the kinetic specificity of the cells in the area and the latter indicates the sensitivity or reactivity of the organisms to the abrupt conditional change (for the relationship of 'responsiveness' and passing ratio, see Appendix).

The arrival ratio is in principle determined by two physiological (kinetic) parameters: the swimming velocity and the frequency of directional change (turning rate or turning frequency). The mean turning rate f_t is given by

$$f_t = \frac{n_t}{N \cdot \Delta t} \, [\text{Hz}]$$

where n_t, N and Δt are the total count of turnings and the total number of cell swimming paths in the superimposed picture and the time range, respectively.

Measuring the mean swimming velocity v can be tedious, although it is given simply by

$$v = \frac{1}{N \cdot \Delta t} \, \Sigma l \, [\mu\text{m/s}]$$

where l is the path length of each swim. To obtain significant numbers of swimming velocity data, a computerized evaluation system is very helpful. (For the relationship between these parameters and the arrival ratio, see Appendix.)

9.4 DETAILED SINGLE CELL TRACKING AND MOTILITY ANALYSIS

For the behavioural analysis of microorganisms based on their motility, such as ciliary or flagellar activity, and even on their physiological conditions such as the membrane potential, detailed single cell tracking is required. There is already microcomputer software developed for analysing a variety of swimming paths in ciliate protozoa (Russo *et al.*, 1988). In this section the mathematical analysis of a ciliate organism *Paramecium caudatum* is discussed as an example.

The setup and method for recording the swimming path is no different from that for the analysis of phototaxis in *P. bursaria*, except for the rather higher magnification so that rough cell shape and its orientation can be recognized. There are several levels of swimming path analysis.

9.4.1. Straightness of the swimming path

If the measurement of the swimming path length (L_p) of a microorganism for a long period (t) is (semi-)automated, obtaining the 'straightness' of the path

$$s = \frac{L_p(t)}{L(t)}$$

is useful, where $L(t)$ is the migration distance reached during the period t, for the manner of swimming depends strongly on the physiological conditions of unicellular organisms. If the measuring period t is long enough compared to the mean free time $t_f (t \gg t_f)$, the mean turning rate (f_t) is estimated as $f_t = s^2$. Otherwise, instead of L_p, the mean swimming velocity v can be used to obtain s, as

$$s = \frac{v \cdot t}{L(t)}$$

The frequency of the spontaneous avoiding reaction f_t is a good indicator of the membrane condition of ciliate and flagellate organisms (Eckert, 1972), since their motile organelles – cilia and flagella – are regulated by the electrical phenomena of the cell membrane, mediated by voltage-sensitive calcium channels on the membrane. Even when the membrane potential of the cell is only a few mV higher than its normal resting level, i.e. when the membrane is slightly depolarized, f_t would increase dramatically. Generally, f_t becomes smaller when the membrane is hyperpolarized. f_t or s is therefore a highly important parameter determining the kinetic behaviour, and is often measured (e.g. Tominaga and Naitoh, 1992, for thermotaxis in *Paramecium*) or qualitatively observed (e.g. Maier and Müller, 1990, for chemotaxis in *Laminaria* spermatozoid) to discuss the mechanism of tactic behaviour.

9.4.2 Estimation of ciliary activity

The turning rate is principally based on the fluctuation of membrane potential and a fairly long period of cell tracking is required to evaluate it. Although the excitability of the cell is clearly indicated by f_t relative to the resting condition, the level of membrane potential in a short period of time, for example between two subsequent spontaneous avoiding reactions during forward swimming, is not to be detected in this method, even qualitatively.

In the cylindrical ciliates, such as *Paramecium*, *Didinium*, *Tetrahymena* or *Loxodes*, some detailed swimming path analyses are available to estimate the momentary membrane conditions. The swimming path of such ciliates is well known as a good indicator of membrane condition (Machemer and Sugino, 1989). Cells of this kind usually swim in a spiral fashion. Their swimming path is in general a result of the combination of three components of motion: forward movement, cell rotation and deflexion of cell orientation, causing a steady change in swimming direction (Naitoh and Sugino, 1984). Not only the forward propulsion but also the other motion components are due to the force generated by the numerous cilia on the cell surface. The cell's rotation around its swimming axis – normally the same as the longitudinal axis – is caused by the difference in direction of the effective stroke of the cilia from that of the cell axis. The deflection occurs when the cell shape or the distribution of ciliary activity on the cell surface is slightly asymmetrical (Fig. 9.8). Thus the spiral swimming movement of a ciliate is the direct expression of ciliary activity, i.e. the beat frequency and beat direction (Sugino and Naitoh, 1988). The ciliary activity then directly indicates the membrane conditions.

The path of a ciliate organism swimming along a horizontal spiral axis is observed as a sinusoidal curve (Fig. 9.9). If the cell has been traced successfully for at least one cycle of the helix of its swimming path, one can obtain the parameters describing the helix, from which the important characteristics of ciliary beating are estimated. Parameters to read from the swimming path are the wave length (λ), duration (t_e) and amplitude ($2a$) of one cycle of the sinusoidal path and the radius (r) of the cell, regarded as a cylinder.

At first the velocity components corresponding to three motions mentioned are derived. The velocity of forward swimming is given by

$$V_f = \pm \frac{\sqrt{\lambda^2 + 4\pi^2 a^2}}{t_e}$$

where the sign is determined by the swimming direction, i.e. forward or backward. For forward swimming, a positive sign should be given. The angular velocity of cell rotation is given by

$$\omega_s = \pm \frac{2\pi\lambda}{\sqrt{\lambda^2 + 2\pi^2 a^2 t_e}}$$

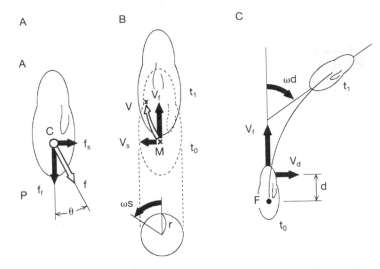

Figure 9.8 Propulsive force and cell movement in a cylindrical ciliate cell. **A**: the propulsive force vector generated by a cilium. When the beat direction is shifted from the direction of the cell axis by as much as angle θ, the propulsive force vector f is dissolved into two vector components on the cell surface, parallel (f_f) and normal (f_s) to the cell axis. Thus $f = f_f + f_s$ and

$$\begin{cases} f_f = f \cos \theta \\ f_s = f \sin \theta \end{cases}$$

where

$$\begin{cases} f = |f| \\ f_f = |f_f| \\ f_s = |f_s|. \end{cases}$$

A: cell anterior; **C**: cilium generating the force f; **P**: cell posterior. **B**: translocation of a marker (M) on the cell surface. Because the cilia distribute more or less uniformly on the cell surface and their beat direction is thought to be synchronously controlled by the membrane potential, the total summation of force ω_s for all cilia tends to rotate the cell around its own axis. r: radius of the cell; V: velocity vector of translocation of M. V is generated as a counteraction against f; V_f: velocity vector of forward movement of the cell corresponding to the component of V parallel to the cell axis; V_s: the component of V normal to the cell axis; ω_s: angular velocity of cell rotation caused by V_s and defined as

$$\omega_s = \frac{V_s}{r}$$

where the sign is determined by the direction of the spiral. When the rotation is counterclockwise, the sign is positive. The angular velocity of deflection of cell orientation is given by

$$\omega_d = \frac{4\pi^2 a}{\sqrt{\lambda^2 + 4\pi^2 a^2 t_e}}$$

The desired physiological characteristics for estimating cell condition, i.e. membrane potential, are the beat frequency and the beat direction (θ) of the cilia. Although θ is correctly the 'net stroke force direction', because the observed propulsive force f also includes the small force component in the opposite direction generated by the recovery stroke, one can regard it as the direction of the effective stroke. Unfortunately, the beat frequency is not given directly from this process, but only the velocity (V) of the movement of points on the cell surface. This parameter is thought to be more or less proportional to the beat frequency in normal physiological conditions. If a voltage clamp is available to measure the membrane potential against the translocation velocity, the membrane potential can be evaluated indirectly but quantitatively.

V as an indicator of ciliary activity is derived as

$$V = \sqrt{V_f^2 + V_s^2}$$

$$= \frac{1}{t_e} \sqrt{\frac{\lambda^4 + 4\pi^2(r^2 + 2a^2)\lambda^2 + 16\pi^4 a^4}{\lambda^2 + 4\pi^2 a^2}}$$

$$= \frac{1}{t_e} \sqrt{L^2 + \left(\frac{2\pi r \lambda}{L}\right)^2}$$

Figure 9.8—*continued from previous page*

where $V_s = |V_s|$. **C**: deflection of cell orientation. When the cell is given any localized force (f_d) acting on it perpendicularly, it tends to change its orientation. Because of this deflection (ω_d), the microorganism cannot swim straight. ω_d is defined as

$$\omega_d = \frac{V_d}{d}$$

where $V_d = |V_d|$. V_d is the velocity vector of the deflection movement at the action point of f_d and d is the distance between the action point and the fulcrum F, around which the deflection occurs. Combined with the rotation ω_s, it makes the swimming path helical. V_f: velocity vector of forward movement.

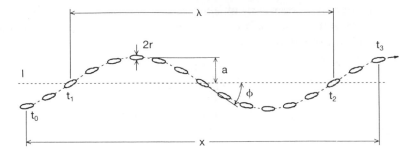

Figure 9.9 Information for estimating ciliary activity available from the swimming path recording. *a*: half value of amplitude of the sinusoidal curve fit to the swimming path; *λ*: wavelength of the sinusoidal curve fit to the swimming path; *r*: radius of the organism regarded as a cylinder; t_0, t_3: time points at the beginning and end of swimming path recording; t_1, t_2: time points at the beginning and end of one cycle of spiral swimming. The duration of one helical swimming cycle t_e is then given by $t_e = t_2 - t_1$; *x*: migration distance during the recording period; *φ*: the slope of the swimming path helix.

where

$$L = \sqrt{\lambda^2 + (2\pi a)^2}$$

which corresponds to the three-dimensional swimming path length of the organism during one cycle of the helix. If *a* is large enough compared to *r*, *V* is approximated to the swimming velocity of the cell observed precisely:

$$V \simeq v = \frac{L}{t_e}$$

The ciliary beat direction *θ* is estimated as

$$\theta = \arctan \frac{V_s}{V_f}$$

$$= \arctan \frac{2\pi r \lambda}{L^2}$$

In the case of *Paramecium*, the relationship between membrane potential and ciliary activity parameters (beat frequency and direction) is precisely investigated electrophysiologically using the voltage-clamp technique (Machemer, 1974, 1976, 1977, 1986; Machemer and Eckert, 1975). Comparing *θ* or *V* with the data of those electrophysiological studies the membrane condition or the ciliary activity of free-swimming cells is thus easily estimated.

9.4.3 Instantaneous ciliary activity

The analytical methods introduced in the preceding sections are very useful in considering the kinetic response of a ciliate organism in terms of membrane potential. On the other hand, in order to analyse precisely the degree of phobic response or change in condition of an organism during its unstable swimming, in relation to the membrane excitation and the temporal change in the membrane potential, the organism must be observed in detail, usually about 30–100 times for organisms of 100–300 μm in length. This detail is needed to measure the cell rotation traced with marker(s) such as the oral apparatus or, if possible, microbead particle(s) applied to the cell surface.

In the analytical process two subsequent images of a cell are compared. The time course of the cell's condition is obtained by alternating comparisons of the two images. The numerical information to be gleaned from the video pictures are the apparent length L_a of the organism along the axis of its rotation, which is normally almost the same as the forward propelling axis or its morphological longitudinal axis; the apparent angle θ_h, i.e. the horizontal component of the solid angle between the axes of two subsequent cell images; the position d_M of a marker M on the cell surface, given as the apparent distance from the cell axis; and the apparent distance y_a of the cell propulsion as the component along the axis of the last cell image (Fig. 9.10). These parameters vary from frame to frame. Additionally, two more constant parameters are required: the radius (r) and the real length (L) of the cell.

L cannot be estimated from only two subsequent cell images, but might be measured as the length of the longest image among many recorded images of the same cell. Also, if a horizontal path of stable swimming of the cell has been recorded for at least one cycle of the helix, L can be estimated using the apparent length (L_a) during stable swimming and the slope (Fig. 9.9, ϕ) of the swimming path helix by the equation

$$L = \frac{L_a}{\cos \phi} \quad .$$

From these parameters the angular velocity of cell rotation ω_s, the angular velocity of deflection of the cell orientation ω_d and the forward propulsion velocity V_f are estimated.

$$\omega_{si} = \frac{\theta_{Mi+1} - \theta_{Mi}}{\Delta t}$$

$$\omega_{di} = \frac{\theta_{di}}{\Delta t}$$

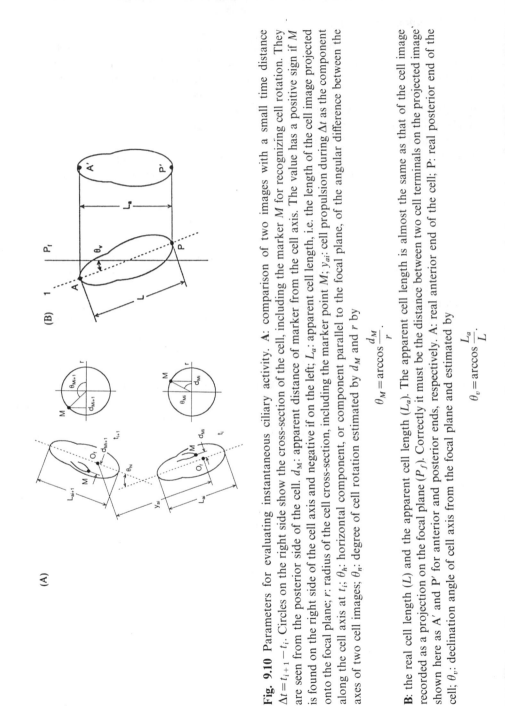

Fig. 9.10 Parameters for evaluating instantaneous ciliary activity. **A**: comparison of two images with a small time distance $\Delta t = t_{i+1} - t_i$. Circles on the right side show the cross-section of the cell, including the marker M for recognizing cell rotation. They are seen from the posterior side of the cell. d_M: apparent distance of marker from the cell axis. The value has a positive sign if M is found on the right side of the cell axis and negative if on the left; L_a: apparent cell length, i.e. the length of the cell image projected onto the focal plane; r: radius of the cell cross-section, including the marker point M; y_{ai}: cell propulsion during Δt as the component along the cell axis at t_i; θ_h: horizontal component, or component parallel to the focal plane, of the angular difference between the axes of two cell images; θ_h: degree of cell rotation estimated by d_M and r by

$$\theta_M = \arccos \frac{d_M}{r}.$$

B: the real cell length (L) and the apparent cell length (L_a). The apparent cell length is almost the same as that of the cell image recorded as a projection on the focal plane (P_f). Correctly it must be the distance between two cell terminals on the projected image shown here as A' and P' for anterior and posterior ends, respectively. A: real anterior end of the cell; P: real posterior end of the cell; θ_v: declination angle of cell axis from the focal plane and estimated by

$$\theta_v = \arccos \frac{L_a}{L}.$$

and

$$V_{fi} = \frac{\omega_i^3 y_i}{\omega_{di} \sin(\omega_i \Delta t) + \omega_i \omega_{si}^2 \Delta t}$$

where

$$\omega_i = \sqrt{\omega_{si}^2 + \omega_{di}^2}$$

θ_{di} is here the real deflection angular velocity given by

$$\theta_{di} = \arccos(\cos \theta_{vi} \cdot \cos \theta_{vi+1} \cdot \cos \theta_{hi} + \sin \theta_{vi} \cdot \sin \theta_{vi+1})$$

(for detail, see Appendix). y_i is measured as the magnitude of the vectoral component of the cell propulsion during Δt parallel to the cell axis at t_i and calculated using

$$y_i = \frac{y_{ai}}{\cos \theta_{vi}}$$

Thus the direction and the activity of ciliary beating at the moment t_i are estimated as

$$\theta_i = \arctan \frac{r \cdot \omega_{si}}{V_{fi}}$$

and

$$V_i = \sqrt{V_{fi}^2 + V_{si}^2}$$

9.4.4 Three-dimensional cell tracking

Utilizing the synchronous recording feature of two video cameras makes three-dimensional tracking of swimming microorganisms possible. A convenient method for recording two views synchronously is to superimpose them on a common frame, applying independent colour signals to each view (Baba *et al.*, 1991; Fig. 9.11). To permit this each view must be taken in monochrome, but the spatial precision of the images is retained and it is the same as that of a single-view recording. Even without an encoder superimposed recording is possible by optically mixing two views illuminated in different colours or coloured by different filters (Izumi-Kurotani *et al.*, 1989).

9.5 VIDEO RECORDING OF RAPID ORGANELLAR MOTION

The high motility of fast-swimming unicellular organisms is due to their motile organelles such as cilia and/or flagella. Since these are very small, the object must be highly magnified for their movement to be observed or

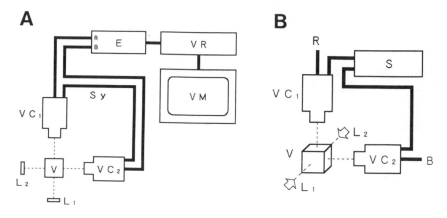

Figure 9.11 A: a connection diagram of the three-dimensional swimming path recording system. Two video signals are combined through a video encoder (E) into a superimposed image. **B**: input channel for blue signal; L_1, L_2: LEDs (light-emitting diodes) for illumination. In this example transparential illumination is applied; R: input channel for red signal; Sy: cable for synchronization signal; V: experimental vessel; VC_1, VC_2: two synchronized video cameras; VM: video monitor; VR: video recorder; (modified from Baba *et al.*, 1991). **B**: with this signal separation method, a single light source (L_1) or a coaxial light source (L_1 and L_2) can be used for the dark-field illumination. If the object microorganism is so small that an individual cell image consists of only a few pixels, two video cameras (VC_1 and VC_2) should be synchronized externally using a colour signal generator (S), in order to avoid the fading out or horizontal shifting of colour information.

recorded. Under such viewing conditions, tracking and catching a free-swimming cell is very difficult. In particular, the motion of these organelles is normally too fast to record as sharp images using a conventional video recorder. A high-speed video recording under a microscope with high magnification is therefore needed. In most cases strobe illumination (Baba and Mogami, 1987) and cell fixation (holding) techniques are applied, and this is the most obvious difference from the other recording techniques for micro-organisms.

9.5.1 Recording of flagellar movement of free-swimming microorganisms

To observe the waveform of flagella of 50–100 μm long, a magnification of $\times$ 400 is usually chosen. With such magnification it is almost impossible to track the same organism for a long time by moving the microscope stage manually. If the object organism species is easy to culture and to gather in one place, and if it is not necessary to observe the same cell continuously, it is worth recording the free-swimming cells.

The camera device of the high-speed video system is mounted on the microscope using a mounting adapter as usual, or hung above it with a shield around its light path and adjusted separately to focus (Fig. 9.12).

The observation vessel can be a combination of a conventional glass slide and a cover glass. In the case of cells with long flagella, the swimming space should be deep enough not to restrict flagellar movement (i.e. about three times the length of the cell, including its flagella). The focal plane should then

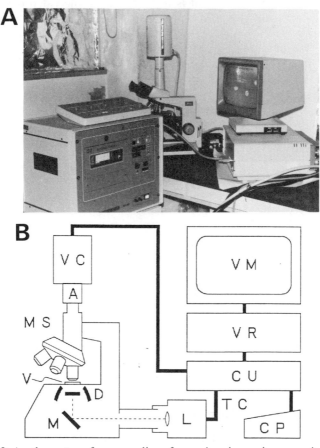

Figure 9.12 A: the setup for recording free-swimming microorganisms with a high-speed video system (NAC HMS-200). Strobe illumination is used for the recording, with high magnification. The frame rate is 200 fps and the light emission of the strobe is triggered by the video system synchronous to the exposure (image input). **B**: the connection diagram of the high-speed video recording system: A: video camera adapter; CP: control panel of video system; CU: control unit of video system; D: dark-field illumination device, e.g. paraboloid condenser; L: strobe illumination; M: mirror, MS: microscope, TC: trigger cable; V: cell observation vessel (for detail, see Fig. 9. 13); VC: video camera; VM: video monitor; VR: video recorder.

be set to the middle level of the swimming space. Fig. 9.13 shows the observation vessel with a spacer to keep the height of the swimming space. Fig. 9.14 shows some examples of flagellated microorganisms recorded with a high-speed video recording system.

If the cell suspension in the vessel is not dense enough, one should search for and track the cell manually. This results in a more natural flagellar movement than if the cell has been restrained. This is very important if the analysis of flagellar movement is to be based on the video pictures.

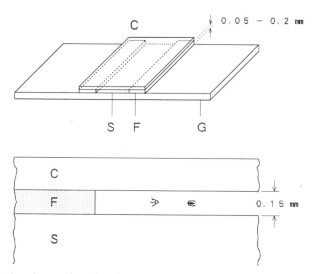

Figure 9.13 An observation chamber typically used for the swimming cell recording. The depth of the swimming area or the suspension space varies from 0.05 to 0.2 mm, depending on the size of the object organism. Since the movement of motile organelles (flagella, tentacle, haptonema, etc.) should not be restricted mechanically by the glass slide or cover glass, the swimming area is recommended to be kept wide enough for the size of the organism, including its motile organelles.

9.5.2 Analysis of flagellar movement

In most cases, the shape of flagellar movement is analysed to investigate the beating mechanism of cilia and flagella based on the sliding of axonemal fibres (Satir, 1965; Brokaw, 1965, 1990). Data on flagellar shape on the video frame are usually stored as a series of x-y coordinate values of points on the flagellar axis. Computerized systems for flagellar axis tracing have already been developed for cinematographic pictures (e.g. Baba and Mogami, 1985) and they are also suitable for handling video images. Since the spatial precision of the video image is normally worse than that of cine film, the picture should be taken as large as possible. In addition, most high-speed video systems utilize picture fields that interlace as an independent frame to compensate for the high frame rate. This reduces the vertical precision of the image to half of that of a conventional one. What should be done carefully in image analysis is the relocation or the location check of the reference point, which is usually shifted one scanning line from frame to frame.

The coordinate data (x, y) of flagellar segments, describing their shape, are at first differentiated along the flagellar axis s to obtain the bending angle θ_b as

$$\theta_b(s) = \arctan \frac{y'(s)}{x'(s)}$$

θ_b is theoretically proportional to the sliding amount (Δl)

$$\Delta l = a\theta_b$$

where a is the distance between adjacent axonemal filaments (doublet microtubules). θ_b is hence also called the shear angle, and is very important in analysing flagellar bending mechanisms (Sleigh, 1974; Brokaw, 1979, 1983; Brokaw and Luck, 1985; Brokaw and Kamiya, 1987).

To standardize θ_b, the coordinate system must be fixed to the orientation of the cell body for each frame. This operation is necessary when θ_b is compared temporally to analyse the sliding velocity of each segment.

When θ_b is again differentiated along the flagellar axis, the curvature is derived for each segment as

$$\frac{1}{r} = \frac{d\theta_b}{ds}$$

The curvature indicates the degree of bending as a result of axonemal sliding (Brokaw, 1985, 1990). By mapping the curvature along the flagellar axis, many features of flagellar movement of the object organism, such as the initiation and the propagation of bending, become visible (Fig. 9.15). Hori *et al.* and the author have been analysing the flagellar movement in many species of unicellular algae to qualitatively characterize the specific beating pattern in each species.

9.5.3 Cell holding for video recording

For long-lasting observation of motile organelles, or for the recording of experimental results requiring any kind of cell manipulation, the object organism must be held somehow to keep it in sight during the recording or experiment, even though mechanical cell holding does sometimes unavoidably influence the organellar motion. In general, the longer, the rarer and the thinner the organelle is, and the faster it moves, the stronger it will be influenced hydrodynamically. In fact, the organism with a single flagellum, for example spermatozoa, is less resistant to hydrodynamic effect than the

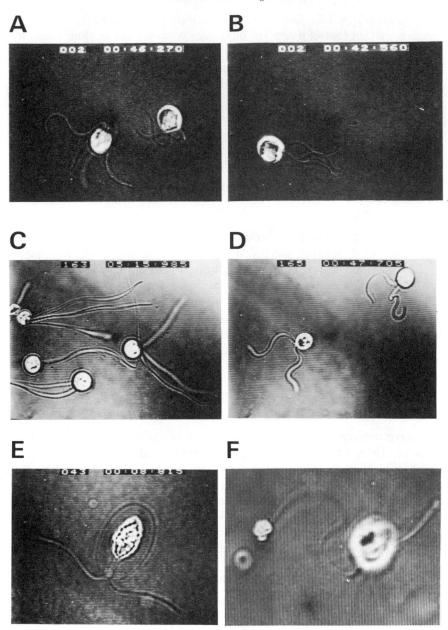

Figure 9.14 Examples of high-speed video pictures of unicellular algae. **A**: the forward swimming of *Cymbomonas tetramitiformis* (Prasinophyceae) in the quadriflagellate stage. Four radial symmetric flagella beat synchronously (× 400). **B**: the backward swimming of *C. tetramitiformis* (Throndsen, 1988) (× 400). **C**: *Pterosperma cristatum* (Prasinophyceae). During swimming, four flagella beat synchronously as a bundle, as in animal spermatozoa (Inouye *et al.*, 1990). This species swims only

cilia surrounding the cell surface of holotrich protozoa (*Paramecium*, *Loxodes*, etc.) or of *Opalina* or the cirri of hypotrich protozoa (*Stylonychia*, *Euplotes*, *Oxytricha*, etc.) or the compound cilia of mussel gills, which are several μm in diameter. There are various studies considering this weak point of sperm flagella (Gibbons *et al.*, 1987; Eshel and Gibbons, 1989; Shingyoji *et al.*, 1991).

There are a number of techniques for single cell holding. One of the easiest methods is to use a kind of glue such as Polylysine or concanavalin A (conA). The glue is first applied to the surface of the glass slide or cover glass, and the object microorganisms are then poured on with the medium. This is, however, often not sufficiently powerful: there are many species strong enough to escape from the trap of these adhesives, or even if trapped, they can easily wander. At worst the motile organelles can adhere to the glass surface and become immotile, rendering the study of their movement impossible.

When the cell body is to be observed and not the motile organelle, gelatine, agarose or methylcellulose can be used to immobilize or to slow down the object microorganism (e.g. Ettienne, 1970; Reize and Melkonian, 1989).

Two other techniques employed to observe ciliary or flagellar movements are the suction micropipette and the holding needle.

In the case of certain adhesive or sedentary microorganisms no special technique is needed (e.g. Hou and Brücke, 1931; Wood, 1970).

Cell holding with a suction pipette

When the cell body is fairly hard, for example surrounded by a cell wall, whether the organism can swim and the cells can be collected densely enough in a suspension medium, or whether the organism is immotile, the microsuction pipette is a very useful tool for tight cell holding. Movement of the organelle is kept mechanically unrestricted (Fig. 9.16) (Rüffer and Nultsh, 1987).

Figure 9.14—*continued from previous page*

backward and instead of the avoiding reaction it stops swimming (× 400). **D**: *Nephroselmis astigmatica* (Prasinophyceae) swimming forward (left) and changing its swimming direction (right) (Inouye and Pienaar, 1984) (× 400). **E**: *Sphaerellopsis* sp. (Chlorophyceae) swimming forward with breast stroke (ciliary type beating) of flagella (× 1000) **F**: *Chrysochromulina hirta* (Prymnesiophyceae). A haptonema accumulating the food substances on its tip beginning to wind to transport them to the posterior end of the cell (Kawachi *et al.*, 1991) (× 500). (Pictures **A–E** taken by I. Inouye, and **F** by M. Kawachi.)

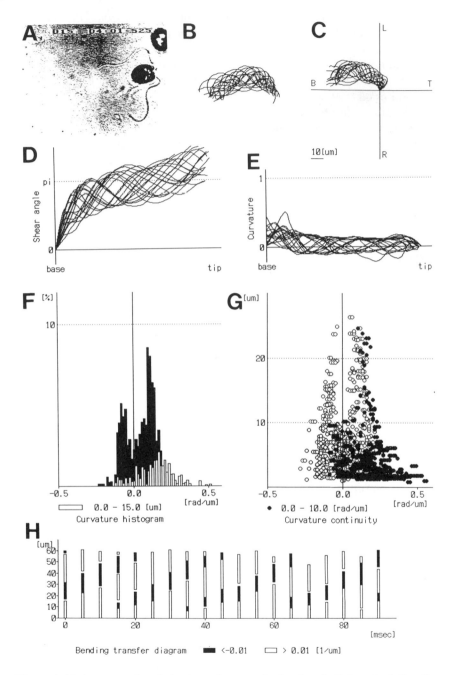

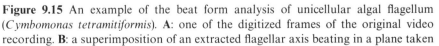

Figure 9.15 An example of the beat form analysis of unicellular algal flagellum (*Cymbomonas tetramitiformis*). **A**: one of the digitized frames of the original video recording. **B**: a superimposition of an extracted flagellar axis beating in a plane taken

Cells are suspended fairly densely in the medium and placed into the cavity of the observation vessel on the microscope stage. The microsuction pipette, with a tip diameter of about half the cell size, is set below the cover glass and in sight of the microscope. By pulling the piston of the syringe or lowering the level of the water reservoir connected to the pipette, negative pressure is generated in the pipette to suck the medium through the narrow opening, and eventually an organism will be trapped at the tip. When the orientation of the cell is not suitable for observation, one can reorient it by lowering the suction pressure temporarily so that the cell can rotate without escaping. Most studies on the flagellar movement of held spermatozoa are performed using this holding technique. The suction pipette is of great use also in electrophysiology (Nichols and Rikmenspoel, 1977; Moreton and Amos, 1979) and for cytochemistry in relation to cell motility (see Fig. 9.19A).

Cell holding with a holding needle

When the number of cells available is limited, or when the cell is not suitable for a suction pipette, a holding needle should be used (Fig. 9.17; Kinosita *et al.*, 1965). This method is used for electrophysiological studies in protozoa (Okajima, 1953; Naitoh, 1959, 1966; Eckert and Sibaoka, 1967; de Peyer and Machemer, 1977). Unfortunately one has to penetrate or press the organism with a finely tapered glass needle, which might cause cell damage. This technique is called the 'hanging drop method', because the selected object cell is at first hung down with a drop of medium below the cover glass (Fig. 9.18). After the cell is completely held with the needle the chamber is filled with the medium to prevent it from drying up. If the drop is fairly

---

Figure 9.15—*continued from previous page*

with a frame rate of 200 Hz for 90 ms. **C**: reorientation of flagellar shapes. Flagellar base is set to the origin and the direction of the *x*-axis corresponds to the cell axis, of which the positive direction is set to the direction of cell anterior. **D**: shear angle diagram. The abscissa indicates the distance from the base along the flagellar axis. **E**: curvature diagram. The bending curvature is plotted along the flagellar axis in rad/μm. Noise is cut through a running filter. **F**: curvature histogram. The occurrence is plotted in percent against curvature. Filled columns: whole flagellar axis; open columns: proximal part. **G**: curvature continuity diagram. The ordinate indicates the length of the fragment with the curvature indicated by the abscissa ±0.05 rad/μm continuously. Open circles: whole flagellar axis; filled circles: proximal part. **H**: bending propulsion time course. The distribution of bent regions to the opposite directions (filled and open columns) is plotted along the flagellar axis. The ordinate indicates the distance from the base. The pattern of bend propagation is clearly shown.

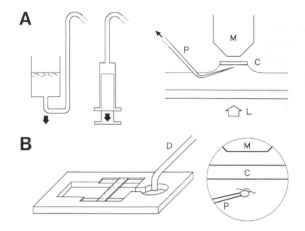

Figure 9.16 A cell-holding chamber with a suction pipette. **A**: a microsuction pipette is set under the microscope (right). Negative pressure is generated by pulling the pipette syringe (middle) or by lowering the water vessel (left). **B**: an observation chamber. Object microorganisms are suspended in the medium in the cavity. The microsuction pipette for cell holding is applied just below the narrow cover glass (C). The tip diameter of the pipette should be suitable for the object species (inset). D: a thick suction pipette for draining the medium; M: microscope.

small it will be easier to catch the cell, but there is a risk that the drop will dry up before cell capture can be achieved. In spite of such difficulties, this technique is often employed because of its reliability in cell holding, especially for vigorous microorganisms such as hypotrich protozoa (Fig. 9.19B; Okajima and Kinosita, 1966; Naitoh and Eckert, 1969; Deitmer *et al.*, 1983; Sugino and Machemer, 1987, 1988, 1990).

9.6 VIDEO RECORDING OF SLOWLY MOVING MICROORGANISMS

In contrast to the fast-swimming microorganisms requiring high-speed video equipment, those with slow movement, such as amoebae, must be observed patiently. Since the frame rate of 30 fps of a normal video recorder is not necessary, the time-lapse recording technique is often employed (e.g. Nelson *et al.*, 1982; Dworkin, 1983).

Rough analysis of amoeboid movement in terms of migration velocity and direction is more or less the same as that for the cell tracking of ciliate organisms, except for the scale of Δt (e.g. Häder and Poff, 1979). To analyse amoeboid movement more precisely – for example the generation and elongation of pseudopodia – the frame comparison technique is useful (Fig. 9.20). For this

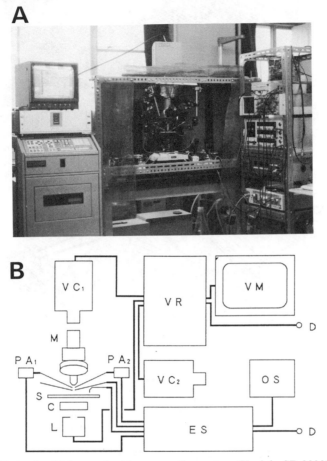

Figure 9.17 A: the high-speed video recording setup (Kodak SP-2000) for electrophysiological experiments with protozoan cells. The setup is foiled with wire gauze in order to shield the microelectrodes from electric noise from the surrounding environment. B: its connection diagram. C: condenser lens; D: data output port to computer; ES: electrophysiological equipment consisting of digital timer, oscilloscope, differential amplifier, voltage/current clamp amplifier, function generator, etc.; L: light source. When a strobe illumination is used it must be synchronized with the video camera; M: microscope with an objective lens of $\times 16 - \times 40$; OS: oscilloscope to monitor the membrane potential, membrane current or the other signals; PA_1, PA_2: preamplifiers for the intra- and extracellular electrodes; S: stage of the microscope, to which the micromanipulators holding the electrodes must be fixed; VC_1: video camera recording the microscopic image; VC_2: video camera recording the membrane potential and the other signals to be recorded, displayed together on the oscilloscope as points moving only vertically. This is recorded together with the microscopic image through VC_1 as an inset view; VM: video monitor; VR: video recorder.

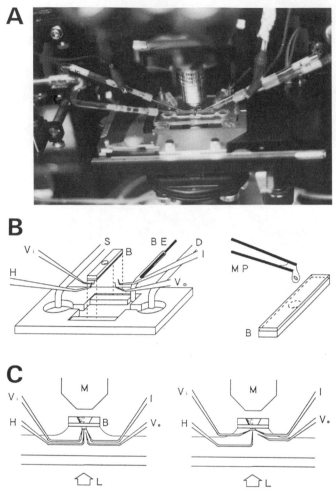

Figure 9.18 **A**: the chamber with a holding microneedle (left) for electrophysiological experiments with protozoan cells. **B**: its schematic diagram (left). **B**: 'bridge', under which the object microorganism cell hangs in a drop of medium (right). The bridge is a tiny metal tip about 1 mm thick with a hole at the centre lined with a thin glass plate. The cell is observed through the hole; BE: bath electrode; D: pipe for draining the medium; H: cell-holding microneedle; I: electrode for current injection; S: pipe to supply medium; V_i: intracellular microelectrode; V_o: extracellular microelectrode called 'reference microelectrode'. All the microneedles have a tip diameter of about 0.2 μm. **C**: tip angle arrangement of the microneedles. The tip of the holding needle to be inserted into or through the cell should be arranged perpendicularly to the glass plate of the bridge. The tip angle of microelectrodes, I and V_i for cylindrical (e.g. *Paramecium*) or globular cells (e.g. *Didinium*) is the same as that of the holding needle (left). On the other hand, the electrodes for flat species (e.g. *Stylonychia*, Fig. 9.19) are recommended to be slanted (right).

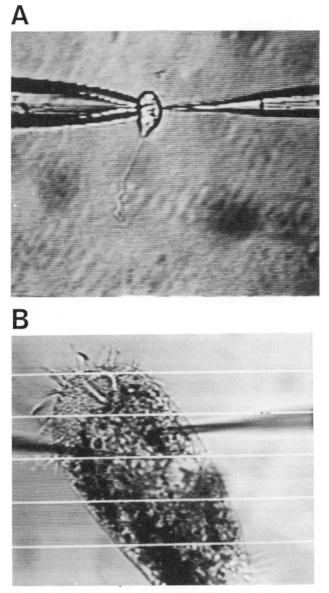

Figure 9.19 Examples of cell holding. **A**: *Vorticella sp.* held with a microsuction pipette (left). In this example this is coupled with microinjection (a needle from the right side) in order to observe the spasmonemal stalk contraction induced by Ca^{2+} ions ($\times 200$). (Experiment and photograph by K. Katoh.) **B**: *Stylonychia lemnae* held with a microneedle for the observation of the voltage-dependent motion of cirri (compound cilia). Two more microneedles (microelectrodes) are applied into the cell in order to control and record the membrane potential. (Experiment and photograph using a high-speed video system by R. Hara.)

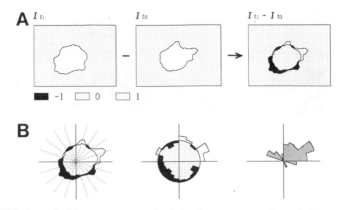

Figure 9.20 Amoeboid movement analysis by frame comparison. **A**: Frame subtraction after binarization. I_{t_0}, I_{t_1}: binarized cell images at two subsequent time points t_0 and t_1. **B**: polarogram of the cell deformation. The balance of cell contour is plotted for each sector around the centre of the cell at t_0. Centre: balance plotting. The circle indicates the zero line and the positive (white) and negative values (black) indicate the generation or elongation of the pseudopod and its contraction, respectively. This diagram shows the distribution of pseudopod generation activity. Right: Translocation tendency. The radius of each fan-out sector indicates the amount of cytoplasmic flow and is plotted as the arithmetic sum of the balances of two opposite sectors. This diagram shows the direction of cell movement and the degree of its centralization.

purpose the video frame must be digitized to feed to the computer (e.g. Turano *et al.*, 1985). On the digitized frame images the cell contour is determined (edge extraction) with the help of appropriate software that recognizes the abrupt change in intensity, i.e. the pixel value, and traces its differential peak. The optics suitable for edge extraction would be dark-field or phase-contrast microscopy. The image of the cell is then binarized by filling the internal area of the extracted cell boundary with '1' and the outside with '0'. The arithmetic subtraction of the binarized frame image recorded Δt before from the target frame shows the new born pseudopodia as the areas of '1' and the cell parts contracted or pulled into the somatic part as the areas of '−1'. A polarogram or radar chart of the change in the shape made by the increase or decrease in the area in each sector indicates the tendency of amoeboid movement.

9.7 SOME NEW VIDEO TECHNIQUES

In contrast to cinematography, the highly sensitive CCD video camera is capable of recording the signal of fluorescence at normal recording speed. Intracellular ion concentration, for example, can be easily measured, and with larger temporal and spatial resolution than with a film camera, cinematography or a photomultiplier. Ca^{2+} concentration in particular is often measured using Ca^{2+}-sensitive fluorescent or luminescent dyes such as Fura-2 or Aequorin (Brundage *et al.*, 1991, in amoebae; Weskampf and Machemer, in preparation, in *Stylonychia*). A membrane potential measurement technique using a voltage-sensitive dye with high temporal and spatial resolution has also recently been developed (Matsumoto and Ichikawa, 1990).

Video recording combined with computerized image processing has enabled the observation of even submicroscopic objects, such as the intracellular organellar motion of mitochondria or microtubules (AVEC-DIC) (Allen *et al.*, 1981; Inoué, 1981; Allen and Allen, 1983). The development of this technique was a great breakthrough for investigators using microorganisms, because most microbiologists had given up the idea of observing motion at the nanometer level.

Another breakthrough might be high-magnification stereomicroscopy (Tieman *et al.*, 1986; Teunis *et al.*, 1992). In particular, Teunis and Machemer are developing high-speed video stereomicroscopy. With this system, together with the author, they are analysing three-dimensional ciliary movements at high time resolution.

Developments in video techniques based on the rapid development of material, memory, sensor and processor make us ambitious to challenge the speed and the complexity of motile microorganisms.

Acknowledgements

The author wishes to thank Professors Y. Naitoh, T. Hori, I. Inouye and Mr Fujimoto for discussions, especially on the analysis of flagellar beating, and Mr M. Kusaka and Dr T. Tominaga for discussions on tactic behaviour analysis. Professor I. Inouye, Drs R. Hara, K. Katoh, M. Kawachi and Mr M. Kusaka have kindly allowed me to use their important pictures. I also thank Professors M.A. Sleigh, S.D. Wratten and H. Machemer, who gave me the chance to introduce video applications in microbiology and for their helpful suggestions, and Ms S. Uesaka for reading the manuscript and for much kind advice.

The author's research discussed herein has been supported by a Grant-in-Aid for the Encouragement of Young Scientists from the Ministry of Education, Science and Culture, and grants from the University of Tsukuba Project Research and from Honda R & D Co., Ltd.

APPENDIX

A1 Cell distribution

Because the rate of change of cell count in an area (e.g. area A) at a certain moment is equal to the net amount of passing rates from each side to the other

$$N'_A(t) = \frac{dN_A(t)}{dt}$$

$$= n_{BA}(t) - n_{AB}(t)$$
$$= T_B N_B(t) - T_A N_A(t)$$
$$= T_B N - (T_A + T_B) N_A(t)$$

Solving the differential equation gives

$$N_A(t) = \alpha e^{-(T_A + T_B)t} + \frac{T_B N}{T_A + T_B}$$

Defining the time constant as

$$\tau = \frac{1}{T_A + T_B}$$

$$N_A(t) = \alpha e^{-t/\tau} + T_B \tau N$$

Since, when $t = 0$ the cell count in area A is $N_A(0) = \alpha + T_B \tau N$, the coefficient α is $\alpha = N_A(0) - T_B \tau N$ and the general equation of cell accumulation then becomes

$$N_A(t) = (N_A(0) - T_B \tau N) e^{-t/\tau} + T_B \tau N$$

In the stationary state after enough time, the cell count in area A is

$$\lim_{t \to \infty} N_A(t) = N_A(\infty) = T_B \tau N$$

and it is known that the cell distribution at equilibrium does not depend on the initial cell distribution ($N_A(0)$ and $N_B(0)$). Hence, the ratio of the equilibrium cell populations in both areas results in

$$\frac{N_B(\infty)}{N_A(\infty)} = \frac{T_A \tau N}{T_B \tau N} = \frac{T_A}{T_B} = I_a$$

which gives the attraction ratio I_a or the attraction index ($\log I_a$) of condition B against condition A. If the population data at the initial state, during the transitional state (t) and in the stationary state are known, the equation

$$N_A(t) = (N_A(0) - N_A(\infty)) e^{-t/\tau} + N_A(\infty)$$

gives the value of τ as

$$\tau = \frac{t}{\log \dfrac{N_A(0) - N_A(\infty)}{N_A(t) - N_A(\infty)}}$$

$$= \frac{1}{\log(N_A(0) - N_A(\infty)) - \log(N_A(t) - N_A(\infty))}$$

From τ and I_a the values of each transfer ratio can be estimated as

$$\begin{cases} T_A = \dfrac{I_a}{(I_a + 1)\tau} \\[3mm] T_B = \dfrac{1}{(I_a + 1)\tau} \end{cases}$$

A2 Relationship between responsiveness and passing ratio

In a simple model where the phobic response occurs immediately after crossing over the boundary and the swimming direction is changed randomly (Fig. A9.1), statistically one half of the cells responding at the boundary (n_{rAB}) will return to the previous area (n_{AA}) and the other half, which have not responded and swum ahead, will cross through the boundary to enter the opposite area (n_{AB}). Thus, the passing ratio from area A to area B (P_{AB}) is written as

$$P_{AB} = \frac{n_{AB}}{n_A}$$

$$= \frac{n_A - n_{AA}}{n_A}$$

$$= 1 - \frac{n_{AA}}{n_A}$$

$$= 1 - \frac{n_{rAB}}{2n_A}$$

$$= 1 - \tfrac{1}{2} R_{AB}$$

where R_{AB} is the responsiveness of cells arriving at the boundary from area A, and is defined as

$$R_{AB} = \frac{n_{rAB}}{n_A}$$

A3 Arrival ratio

Assuming that the cells move in the manner of a random walk, and that the turning rate f_t and swimming velocity v are independent of each other, the mean free path l_f is given by

$$l_f = \frac{v}{f_t}$$

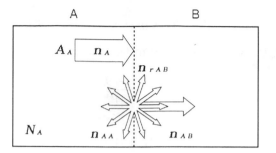

Figure A9.1 Responsiveness and passing ratio. A_A: arrival ratio specific to area A; n_A: number of cells arriving at the boundary in a unit time; n_{AA}: number of cells re-entering area A from the boundary because of their own avoiding reaction; n_{AB}: number of cells entering the new area (B) without or in spite of the avoiding reaction (phobic response); n_{rAB}: number of cells showing the avoiding reaction at the boundary.

If f_t is large enough, the Markov process can be applied and the mean migration velocity L will be approximated to

$$L \simeq l_f \sqrt{f_t}$$

$$= \frac{v}{\sqrt{f_t}}$$

The probability of arriving at or crossing the boundary of an organism existing at a distance x from the boundary during a unit time (Fig. A9.2) is

$$\begin{cases} p(x) = \dfrac{1}{\pi} \arccos \dfrac{x}{L} \ (0 < x < L) \\ p(x) = 0 \ (L < x) \end{cases}$$

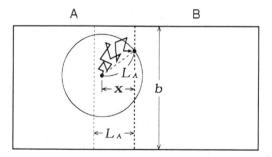

Figure A9.2 Cell migration regarded as a Brownian movement. b: length of the boundary; L_A: mean migration velocity specific for area A; x: a component of the migration distance normal to the line of boundary.

Thus, the total count of cells arriving at the boundary from area A is

$$n_A = \rho_A b \int_0^L p(x)\,\mathrm{d}x$$

$$= \frac{b}{\pi} \rho_A L_A$$

where ρ_A is the cell density in area A, b is the length of the boundary and L_A is the mean migration velocity of cells in area A.

A4 Angular velocity of deflection

When the direction of cell propulsion is defined as the x-axis of the $x-y$ plane corresponding to the focal plane (P_f), the three-dimensional vector of forward swimming V_{fi} at a certain time point t_i is derived from observed angles, as

$$V_{fi} = (V_{fxi}, V_{fyi}, V_{fzi})$$
$$= (V_{fi} \cos \theta_{vi}, 0, V_{fi} \sin \theta_{vi})$$

where $V_{fi} = |V_{fi}|$ (Fig. A9.3). On the other hand, that (V_{fi+1}) at time point t_{i+1} is because of the deflection θ_{hi}:

$$V_{fi+1} = (V_{fxi+1}, V_{fyi+1}, V_{fzi+1})$$
$$= (V_{fi+1} \cos \theta_{vi+1} \cos \theta_{hi}, V_{fi+1} \cos \theta_{vi+1} \sin \theta_{hi} V_{fi+1} \sin \theta_{vi+1}),$$

where $V_{fi+1} = |V_{fi+1}|$.

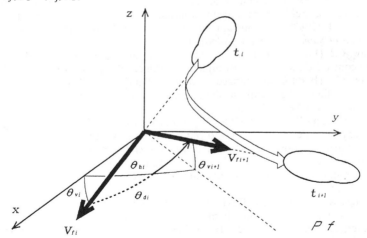

Figure A9.3 Deflection angle (θ_d) of swimming direction. θ_{di} is defined as the angle between the vectors of forward propulsion velocity (V_{fi}, V_{fi+1}) at time points t_i and t_{i+1}; θ_{hi}: accurate deflection angle of cell orientation; $\theta_{vi}, \theta_{vi+1}$: vertical orientations of cell axis at t_i and t_{i+1}.

According to scalar multiplication

$$V_{fi} \cdot V_{fi+1} = V_{fi} V_{fi+1} \cos \theta_{di}$$
$$= V_{fxi} V_{fxi+1} + V_{fyi} V_{fyi+1} + V_{fzi} V_{fzi+1}$$
$$\cos \theta_{di} = \cos \theta_{vi} \cos \theta_{vi+1} \cos \theta_{hi} + \sin \theta_{vi} \sin \theta_{vi+1}$$

is derived.

REFERENCES

Allen, R.D. and Allen, N.S. (1981) Videomicroscopy in the study of protoplasmic streaming and cell movement. *Protoplasma*, **109**, 209–216.

Allen, R.D. and Allen, N.S. (1983) Video-enhanced microscopy with a computer frame memory. *Journal of Microscopy*, **129**(1), 3–17.

Allen, R.D., Allen, N.S. and Travis, J.L. (1981) Video-enhanced contrast, differential interference contrast (AVEC-DIC) microscopy: a new method capable of analyzing microtubule-related motility in the reticulopodial network of *Allogromia laticollaris*. *Cell Motility*, **1**, 291–302.

Baba, S.A., Inomata, S., Ooyama, M. *et al.* (1991) Three-dimensional recording and measurement of swimming paths of microorganisms with two synchronized monochrome cameras. *Review of Scientific Instrumentation*, **62**(2), 540–541.

Baba, S.A. and Mogami, Y. (1985) An approach to digital image analysis of bending shapes of eukaryotic flagella and cilia. *Cell Motility*, **5**, 475–489.

Baba S.A. and Mogami, Y. (1987) Device for controlled intensification of the light output of xenon flash tubes. *Review of Scientific Instrumentation*, **58**(7), 1312–1313.

Brokaw, C.J. (1965) Non-sinusoidal bending waves of sperm flagella. *Journal of Experimental Biology*, **43**, 155–169.

Brokaw, C.J. (1979) Calcium-induced asymmetrical beating of Triton-demembranated sea urchin sperm flagella. *Journal of Cell Biology*, **82**, 401–411.

Brokaw, C.J. (1983) The constant curvature model for flagellar bending patterns. *Journal of Submicroscopic Cytology*, **15**(1), 5–8.

Brokaw, C.J. (1985) Computer simulation of flagellar movement. VI. Simple curvature-controlled models are incompletely specified. *Biophysics Journal*, **48**, 633–642.

Brokaw, C.J. (1990) Computerized analysis of flagellar motility by digitization and fitting of film images with straight segments of equal length. *Cell Motility and Cytoskeleton*, **17**(4), 309–316.

Brokaw, C.J. and Kamiya, R. (1987) Bending patterns of *Chlamydomonas* flagella: IV. Mutants with defects in inner and outer dynein arms indicate differences in outer dynein arm function. *Cell Motility and Cytoskeleton*, **8**, 68–75.

Brokaw, C.J. and Luck, D.J.L. (1985) Bending patterns of *Chlamydomonas* flagella: III. A radial spoke head deficient mutant and a central pair deficient mutant. *Cell Motility*, **5**, 195–208.

Brundage, R.A., Fogarty, K.E., Tuft, R.A. and Fay, F.S. (1991) Calcium gradients underlying polarization and chemotaxis of eosinophils. *Science*, **254**, 703–706.

de Peyer, J.E. and Machemer, H. (1977) Membrane excitability in *Stylonychia*: Properties of the two-peak regenerative Ca-response. *Journal of Comparative Physiology*, **121**, 15–32.

Deitmer, J.W., Machemer, H. and Martinac, B. (1983) Simultaneous recording of responses of membranelles and cirri in *Stylonychia* under membrane voltage-clamp. *Journal of Submicroscpic Cytology*, **15**(1), 285–288.

Donovan, R.M., Goldstein, E., Kim, Y. *et al.* (1986) A quantitative method for the analysis of cell shape and locomotion. *Histochemistry*, **84**(4–6), 525–529.

Dusenbery, D.B. (1985) Using a microcomputer and video camera to simultaneous track 25 animals. *Computers in Biology and Medicine* **15**(4), 169–175.

Dworkin, M. (1983) Tactic behavior of *Myxococcus xanthus*. *Journal of Bacteriology*, **154**(1), 452–459.

Eckert, R. (1972) Bioelectric control of ciliary activity. *Science*, **176**, 473–481.

Eckert, R. and Sibaoka, T. (1967) Bioelectric regulation of tentacle movement in a dinoflagellate. *Journal of Experimental Biology*, **47**, 433–466.

Eggersdorfer, B. and Häder, D.-P. (1991) Phototaxis, gravitaxis and vertical migrations in the marine dinoflagellate *Prorocentrum micans*. *FEMS Microbiology and Ecology*, **85**, 319–326.

Eshel, D. and Gibbons, I.R. (1989) External mechanical control of the timing of bend initiation in sea urchin sperm flagella. *Cell Motility and Cytoskeleton*, **14**, 416–423.

Ettienne, E.M. (1970) Control of contractility in *Spirostomum* by dissociated calcium ions. *Journal of General Physiology*, **56**, 168–179.

Gibbons, I.R., Shingyoji, C., Murakami, A. and Takahashi, K. (1987) Spontaneous recovery after experimental manipulation of the plane of beat in sperm flagella. *Nature*, **325**(6102), 351–352.

Häder, D.-P. and Lebert, M. (1985) Real time computer-controlled tracking of multi microorganisms. *Photochemistry and Photobiology*, **42**(5), 509–514.

Häder, D.-P. and Poff, K.L. (1979) Light-induced accumulation of *Dictyosterium discoideum* amoebae. *Photochemistry and Photobiology*, **29**(6), 1157–1162.

Hasegawa, K., Tanakadate, A. and Ishikawa, H. (1988) A method for tracking the locomotion of an isolated microorganism in real time. *Physiology and Behaviour*, **42**(4), 397–400.

Hou, C.L. and Brücke, E. Th. (1931) Reizversuche an Vorticellen (Alles-oder-nichts-Gesatz, Dekrement der Erregungsleitung in der Narkose, Chronaxie). *Pflügers Archiv*, **226**, 411–417.

Inoué, S. (1981) Video image processing greatly enhances contrast, quality, and speed in polarization-based microscopy. *Journal of Cell Biology*, **89**, 346–356.

Inouye, I, and Hori, T. (1991) High-speed video analysis of the flagellar beat and swimming patterns of algae: possible evolutionary trends in green algae. *Protoplasma*; **164**, 54–69.

Inouye, I., Hori, T. and Chihara, M. (1990) Absolute configuration analysis of the flagellar apparatus of *Pterosperma cristatum* (Prasinophyceae) and consideration of its phylogenetic position. *Journal of Phytology*, **26**, 329–344.

Inouye, I. and Pienaar, R.N. (1984) Light and electron microscope observations on *Nephroselmis astigmatica* sp. nov. (Class Prasinophyceae). *Nordic Journal of Botany*, **4**, 409–423.

Iwatsuki, K. and Naitoh, Y. (1981) The role of symbiotic *Chlorella* in photoresponse of *Paramecium bursaria*. *Proceedings of the Japanese Academy*, **57**(B8), 318–323.

Iwatsuki, K. and Naitoh, Y. (1983) Behavioral responses in *Paramecium micronucleatum* to visible light. *Photochemistry and Photobiology*, **37**(4), 415–419.

Izumi-Kurotani, A., Yamashita, M., Mogami, Y. and Baba, S.A. (1989) Three-dimensional swimming behaviour of *Paramecium* under various gravitational environments. *Zoological Science*, **6**(6), 1089.

Katz, D.F. and Overstreet, J.W. (1981) Sperm motility assessment by video-micrography. *Fertility and Sterility*, **35**(2), 188–193.

Kawachi, M., Inouye, I., Maeda, O. and Chihara, M. (1991). The haptonema as a food-capturing device: observation on *Chrysochromulina hirta* (Prymnesiophyceae). *Phycologia*, **30**(6), 563–573.

Kinosita, H., Murakami, A. and Yasuda, M. (1965) Interval between membrane potential change and ciliary reversal in *Paramecium* immersed in Ba-Ca mixture. *Journal of the Faculty of Science of the University of Tokyo*, IV, **10**(3), 421–425.

Machemer, H. (1974) Frequency and directional response of cilia to membrane potential change in *Paramecium. Journal of Comparative Physiology*, **92**, 293–316.

Machemer, H. (1976) Interactions of membrane potential and cations in regulation of ciliary activity in *Paramecium. Journal of Experimental Biology*, **65**, 427–448.

Machemer, H. (1977) Motor activity and bioelectric control of cilia. *Fortschritte der Zoologie*, **24**(2/3), 196–210.

Machemer, H. (1986) Electromotor coupling in cilia. *Fortschritte der Zoologie*, **33**, 205–250.

Machemer, H. and Eckert, R. (1975) Ciliary frequency and orientational responses to clamped voltage step in *Paramecium. Journal of Comparative Physiology*, **104**, 247–260.

Machemer, H., Machemer-Röhnisch, S., Bräucker, R. and Takahashi, K. (1991) Gravikinesis in *Paramecium*: theory and isolation of a physiological response to the natural gravity vector. *Journal of Comparative Physiology A*, **168**, 1–12.

Machemer, H. and Sugino, K. (1989) Electrophysiological control of ciliary beating: a basis of motile behaviour in ciliate protozoa. *Comparative Biochemistry and Biophysics*, **94A**, 365–374.

Maier, I. and Müller, D.G. (1990) Chemotaxis in *Laminaria digitata* (Phaeophyceae) I. Analysis of spermatozoid movement. *Journal of Experimental Botany*, **41**(228), 869–876.

Matsumoto, G. and Ichikawa, M. (1990) Optical system for real-time imaging of electrical activity with a 128×128 photopixel array. *Society of Neuroscience Abstracts*, **16**, 490.

Moreton, R.B. and Amos, W.B. (1979) Electrical recording from the contractile ciliate Zoothamnium geniculatum Ayrton. *Journal of Experimental Biology*, **83**, 159–167.

Naitoh, Y. (1959) Relation between the deformation of the cell membrane and the change in beating direction of cilia in *Opalina. Journal of the Faculty of Science of the University of Tokyo, IV*, **8**(3), 357–369.

Naitoh, Y. (1966) Reversal response elicited in nonbeating cilia of *Paramecium* by membrane depolarization. *Science*, **154**(3749), 660–662.

Naitoh, Y. and Eckert, R. (1969) Ciliary orientation: controlled by cell membrane or by intracellular fibrils? *Science*, **166**, 1633–1635.

Naitoh, Y. and Kaneko, H. (1972) Reactivated Triton-extracted models of *Paramecium*: modification of ciliary movement by calcium ions. *Science*, **176**, 523–524.

Naitoh, Y. and Sugino, K. (1984) Ciliary movement and its control in *Paramecium. Journal of Protozoology*, **31**, 31–40.

Nakazato, H. and Naitoh, Y. (1980) Factors determining the chemotactic behaviour in *Paramecium. Zoological Magazine (Tokyo)*, **89**(4), 434.

Nakazato, H. and Naitoh, Y. (1981) Chemokinesis in *Paramecium caudatum. Zoological Magazine (Tokyo)*, **90**(4), 677.

Nakazato, H. and Naitoh, Y. (1982) Chemokineses in *Paramecium* (II) Response to 'attractant' and 'repellant'. *Zoological Magazine (Tokyo)*, **91**(4), 422.

Nelson, G.A., Roberts, T.M. and Ward, S. (1982) *Caenorhabditis elegans* spermatozoan locomotion. Amoeboid movement with almost no actin. *Journal of Cell Biology*, **92**, 121–131.

Nichols, K.M. and Rikmenspoel, R. (1977) Mg^{2+}-dependent electrical control of flagellar activity in *Euglena*. *Journal of Cell Science*, **23**, 211–225.

Okajima, A. and Kinosita, H. (1966) Ciliary activity and coordination in *Euplotes eurystomus* – I. Effect of microdissection of neuromotor fibres. *Comparative Biochemistry and Physiology*, **19**, 115–131.

Okajima, A. (1953) Studies on the metachronal wave in *Opalina* I. Electrical stimulation with the micro-electrode. *Japanese Journal of Zoology*, **11**, 87–100.

Reize, I.B. and Melkonian, M. (1989) A new way to investigate living flagellated/ciliated cells in the light microscope: immobilization of cells in agarose. *Botanica Acta*, **102**(2), 145–151.

Ricci, N. (1990) The behaviour of ciliated protozoa. *Animal Behaviour*, **40**, 1048–1069.

Rüffer, U. and Nultsch, W. (1987) Comparison of the beating of *cis*- and *trans*-flagella of *Chlamydomonas* cell held on micropipettes. *Cell Motility and Cytoskeleton*, **7**, 87–93.

Russo, A., Gualtieri, P. and Ricci, N. (1988) A semiautomatic computerized analysis of tracks of ciliates. *Experientia*, **44**, 277–280.

Satir, P. (1965) Studies on cilia II. Examination of the distal region of the ciliary shaft and the role of the filaments in motility. *Journal of Cell Biology*, **26**, 805–834.

Shingyoji, C., Gibbons, I.R., Murakami, A. and Takahashi, K. (1991) Effect of imposed head vibration on the stability and wave-form of flagellar beating in sea-urchin spermatozoa. *Journal of Experimental Biology*, **156**, 63–80.

Sleigh, M.A. (ed) (1974) *Cilia and Flagella*, Academic Press, London.

Sugino, K. and Machemer, H. (1987) Axial-view recording: an approach to assess the third dimension of the ciliary cycle. *Journal of Theoretical Biology*, **125**, 67–82.

Sugino, K. and Machemer, H. (1988) The ciliary cycle during hyperpolarization-induced activity: an analysis of axonemal function parameters. *Cell Motility and Cytoskeleton*, **11**, 275–290.

Sugino, K. and Machemer, H. (1990) Depolarization controlled parameters of the ciliary cycle and axonemal function. *Cell Motility and Cytoskeleton*, **16**, 251–265.

Sugino, K. and Naitoh, Y. (1988) Estimation of ciliary activity in *Paramecium* from its swimming path. *Seitai no Kagaku*, **39**(5), 485–490.

Tuenis, P.F.M., Bretschneider, F. and Machemer, H. (1992). Real-time three-dimensional tracking of fast-moving microscopic objects. *Journal of Microscopy (Oxford)*, **168**(3), 275–288.

Throndsen, J. (1988) *Cymbomonas* Sciller (Prasinophyceae) reinvestigated by light and electron microscopy. *Archiv für Protistenkunde*, **136**, 327–336.

Tieman, D.G., Murphey, R.K., Schmidt, J.T. and Tieman, S.B. (1986) A computer-assisted video technique for preparing high-resolution pictures and stereograms from thick specimens. *Journal of Neuroscience Methods*, **17**, 231–245.

Tominaga, T. and Naitoh, Y. (1992) Membrane potential responses to thermal stimulation and the control of thermoaccumulation in *Paramecium caudatum*. *Journal of Experimental Biology* (in press).

Turano, T.A., D'Arpa, P., Clark, W.L. and Williams, J.R. (1985) A time-lapse, image digitization videomicroscope system based on a minicomputer with large peripheral memory. *Computers in Biology and Medicine*, **15**(4), 177–185.

204 *Video and microorganisms*

Van Houten, J., Hansma, H. and Kung, C. (1975) Two quantitative assays for chemotaxis in *Paramecium*. *Journal of Comparative Physiology*, **104**, 211–223.

Van Houten, J., Hauser, D.C.R. and Levandowsky, M. (1981) Chemosensory behavior of protozoa, in *Biochemistry and Physiology of Protozoa* 4 (eds M. Levandowsky and S. Hutner), Academic Press, New York, pp. 67–125.

Wood, D.C. (1970) Electrophysiological studies of the protozoan, *Stentor coeruleus*. *Journal of Neurobiology*, **1**(4), 363–377.

Index

Page numbers appearing in **bold** refer to figures and page numbers appearing in *italic* refer to tables.